# PRAISE FOR *IS REMOTE WARFARE MORAL?*

"Brilliantly thought-provoking. Joe Chapa's personal experience employing unmanned aerial vehicles in support of combat operations, and his carefully researched and considered analysis, have produced the most important consideration of this aspect of current and future warfare yet written. As with nuclear weapons and chemical warfare, there are no definitive answers, just agonizing judgments that will be hellishly difficult in the moment of existential conflict. A must-read to understand warfare of the future."

—General (Ret.) Stanley McChrystal

"Joseph Chapa's book is a profound meditation on the ethical dimensions of remote warfare, informed by philosophy, psychology, and his own experience as a remotely-piloted aircraft operator. It unflinchingly engages with issues of life, death, and military honor with uncommon care, candor, and insight."

—Milton Regan Jr., McDevitt Professor of Jurisprudence and codirector, Center on National Security and the Law, Georgetown University

"An incredibly valuable addition to the debate over perhaps the most important new question of life and death in modern war. Chapa brings in much-needed personal and professional experience to weigh the morality of drones and remote killing."

—P.W. Singer, author of *Ghost Fleet* and *Wired for War*

"Chapa's examination of the profound tension between remote warfare and close combat make *Is Remote Warfare Moral?* a book that truly matters for military professionals and the societies they serve. Is remote warfare virtuous, and is the ability to surveil and engage targets remotely legitimate? These are the questions addressed in this important book."

—Norton A. Schwartz, president, Institute for Defense Analyses, and former chief of staff, US Air Force

# IS
# REMOTE
# WARFARE
# MORAL?

WEIGHING ISSUES OF
LIFE + DEATH
FROM 7,000 MILES

# IS
# REMOTE
# WARFARE
# MORAL?

JOSEPH O. CHAPA

PUBLICAFFAIRS

New York

PublicAffairs
Hachette Book Group
1290 Avenue of the Americas, New York, NY 10104
www.publicaffairsbooks.com
@Public_Affairs

Printed in the United States of America

First Edition: July 2022

Published by PublicAffairs, an imprint of Perseus Books, LLC, a subsidiary of Hachette Book Group, Inc. The PublicAffairs name and logo is a trademark of the Hachette Book Group.

The Hachette Speakers Bureau provides a wide range of authors for speaking events. To find out more, go to www.hachettespeakersbureau.com or call (866) 376-6591.

The publisher is not responsible for websites (or their content) that are not owned by the publisher.

Print book interior design by Amy Quinn.

Library of Congress Cataloging-in-Publication Data

Names: Chapa, Joseph O., author.
Title: Is remote warfare moral? : weighing issues of life and death from
    7,000 miles / Joseph O. Chapa.
Description: New York City : PublicAffairs, [2022] | Includes
    bibliographical references and index.
Identifiers: LCCN 2021059957 | ISBN 9781541774452 (hardcover) | ISBN
    9781541774469 (epub)
Subjects: LCSH: Drone warfare--Moral and ethical aspects--United States. |
    Targeted killing--Moral and ethical aspects--United States. | Drone
    warfare--Government policy--United States. |
    Terrorism--Prevention--Government policy--United States.
Classification: LCC UG1242.D7 C445 2022 | DDC 623.74/69--dc23/eng/20220121
LC record available at https://lccn.loc.gov/2021059957

ISBNs: 9781541774452 (hardcover); 9781541774469 (ebook)

LSC-C

Printing 1, 2022

*This book is dedicated to the*
*Airmen flying the line.*

# CONTENTS

# INTRODUCTION

## KILLING FROM 7,000 MILES

Eawas called in as the safety observer on a Predator strike against a senior al Qaeda commander in Afghanistan. The Predator remotely piloted aircraft circled thousands of feet above the rugged Afghan terrain while its pilot, its crew, and I remained seven thousand miles to the west in the continental United States. The pilot, Dave Brown, was a newly minted lieutenant with only about a month of Predator experience under his belt.[1] I was older, a captain, and an instructor pilot in the aircraft. One significant benefit of taking the cockpit out of the airplane and putting it on the ground is that we can add some experience or an extra set of eyes to a mission without having to land the airplane to change out the crews. My job as safety observer was to watch and listen, to be prepared to offer help and support if the crew needed it, and otherwise to keep my mouth shut.

The special operations team on the ground wanted Dave and his crew to strike the al Qaeda commander under his crosshairs with a Hellfire missile. This enemy commander—a "high-value individual"

1

for whom the team and numerous Predator crews had been searching for weeks—was finally out in the open. Dave watched the screen and keyed the mic. "Standby. Standby," he said. "There are kids in the field of view. Confirm you copy kids?" Dave continued to watch as the target's children fluttered around him on the silent video monitor.

The radio was quiet for a few moments while the attack controller conferred with the ground force commander. The question the ground team was discussing was one of proportionality. The proportionality requirement in just war theory permits foreseeably killing noncombatants as a side effect only if the moral good to be achieved by killing them significantly outweighs the moral costs.[2] Was this al Qaeda commander of sufficient military value that the good to be achieved could justify foreseeably killing two children? He was out in the open. Dave, the special operations team, and I all knew that if the crew did not strike him today, they might never get another chance. Eventually, the controller on the ground responded. "I copy kids. I see the kids. But when I tell you to shoot, you're gonna shoot."

Lieutenant Dave was asked to do a grave and, frankly, terrible thing that night. Most just war theorists agree that the al Qaeda commander had given up his right not to be killed but that the two children had done nothing to give up their rights. Dave was asked to kill the al Qaeda leader, knowing full well that in doing so, he would also kill children. He was asked to weigh the military objective on one side of the scale against the lives of two innocent children on the other. In this crucible of moral conflict, the twenty-three-year-old lieutenant had to adjudicate between competing virtues of justice, loyalty, and mercy—and the conflicting duties to defend against unjust attacks and to refrain from harming the innocent. At first glance, it might look as though Dave was faced with the peculiar ethical challenges posed by remote warfare. But in those brief moments, he was forced to grapple not with the ethics of remote warfare in particular but with the ethics of war. Proportionality calculus in war—deciding whether a military objective is worth the foreseeable moral cost—is always fraught, whether one is in a Predator cockpit, a special operations raid, or a conventional infantry patrol. Professional obligations and moral ones regularly come into conflict among warfighters on every battlefield, not just remote warfighters. Dave faced agonizing moral questions that night not because so-called drones are at the fringes of permissible human activity but because war is.

Though Dave was the pilot in command of that aircraft, he was not its sole crew member. An enlisted sensor operator sat to his right, and an intelligence analyst remained on the intercom, though he worked in a different part of the building. Dave, showing considerable wisdom for someone with so little experience, recognized that whatever decision he made, his two crew members would have to live with it too. He queried both. "You guys good with this?" Both responded in the affirmative. Despite the assent of the crew and the ground force's claim that the strike would have met all the relevant legal requirements, Dave keyed the mic and told the special operations team that he was going to wait five minutes. If the kids had not departed by then, he would be willing to take the shot. Within five minutes, the children walked back to the house and we all watched as the al Qaeda commander went for a walk by himself. Dave and his crew went to work, and the al Qaeda commander never returned.

Dave's story is just one among many. It is special in the sense that Dave showed great poise and leadership under stress. But in another sense, it is similar to thousands of other Predator sorties. Like so many others, Dave's experience is insightful because it defies many stereotypes that have grown up around these weapons systems. This was not push-button warfare and the crew did not have a PlayStation mentality. The airplane was neither semiautonomous nor robotic. Instead, professional airmen demonstrated prudence in delaying the strike and moral courage in pushing back against the special operations team's demand.[3] They weighed, on the one hand, their professional and moral obligations to prevent the al Qaeda commander from conducting future attacks and, on the other, their professional and moral obligations to minimize harm to innocent people. Dave recognized the psychological burden that might result from taking the lives of these two children. He further recognized that the burden would not be his alone but would be equally borne by his subordinate crew members.

When most people reflect on remote warfare of this kind—and when they draw ethical conclusions about it—they tend to consider the machine overhead and the people on the ground below. They consider the very real humanity of those on the ground, people like the special operations team, the al Qaeda leader, and the civilians, and the moral costs they bear—and rightfully so. But on the other end of the data links and half a world away, there is a tremendous amount of humanity at work.

These remotely piloted weapons systems have been the subject of intense controversy. With pilots and sensor operators seven thousand miles removed from the heat of battle, remote warfare looks to many like the epitome of superpower impunity. The United States can use violence to achieve its aims around the world without exposing its own military to combat risks. Some suggest that these weapons systems—and especially their use in targeted killing outside war zones—are eroding global conceptions of state sovereignty. Others have argued that the ability to conduct military operations without risk to one's own forces will further insulate the American public from their wars. And still others have argued that this attempt at war without risk undermines the very morality of war. Remote warfare, according to some of these arguments, seems to fall outside the boundaries of just warfare. Even now, two decades after the first Predator missile strike on the opening night of the US war in Afghanistan, these controversies persist. If anything, recent events have caused them to intensify.

In January 2020, the United States used an MQ-9 Reaper to kill Iranian general Qasem Soleimani. The Reaper is the remotely piloted aircraft that replaced the Predator. Soleimani's killing raised all the familiar questions. The crew who released the weapon remained half a world away; the strike took place within Iraq's sovereign borders and without the Iraqi government's consent; and several other people were killed in the attack alongside Soleimani.[4] Does this kind of action count as war? Can it be morally justified under the principles that govern war? And if the United States can target leaders of states with which it is not at war, in this case, Iran, what does that mean about other states' ability or license to target US leaders?

At first, we might think this ability to target enemies remotely is unequivocally immoral. War is a terrible and devastating thing, and we ought to resort to war only when faced with the highest of stakes and when all other efforts have failed. But this ability to kill remotely threatens to make military violence an everyday occurrence. Though war ought to remain a dreadful last resort, remote weapons threaten to make warfare commonplace.

Is it plausible that states that have remote weapons will be more likely to resort to violence? Perhaps. But the security challenges the United States has faced since 9/11 are complicated, and questions about the morality of remote warfare are more difficult to answer than they might appear. Imagine that you are president of the United States in the post-9/11 world. Terror networks seek to attack US citizens at home and abroad—first al Qaeda and later its affiliates in the Arabian Peninsula and Iraq. Al Qaeda in Iraq has evolved into the so-called Islamic State (ISIS). The Haqqani network has reconstituted in Afghanistan and Pakistan, and al-Shabaab seeks an Islamic state in Somalia. Those who elected you, and even those who didn't, expect you to do what you can to keep them safe. What will you do?

Well, there is one extreme option. Perhaps you would lay a demand on your fellow heads of state: tell them to prevent the terror networks within their borders from harming US citizens. And if they cannot or will not do that, then you might have to go to war with the terrorists within their borders. If you choose this extreme, you will probably end up starting wars around the globe. The moral costs would be significant and would probably far exceed the benefits.

Or you could retreat to the other extreme. You could choose never to use military force to disrupt or degrade terror networks' capabilities. You might still use diplomacy, intelligence methods, and economic tools to prevent terror networks' most violent attacks, but you would refrain from using military force. If you choose this option, you will undoubtedly accept increased risk to the US citizens whom you are charged with protecting. Here, too, there will be moral costs.

As is so often the case, the best solution probably falls somewhere between these two extremes. As president, you will likely try to refrain from resorting to military force unless it's absolutely necessary. And when it is necessary, you will probably attempt to use as little force as possible to achieve the justified effect. This has been the strategic appeal of remote warfare nearly from its inception.

Four presidents from both parties have employed remote warfare to defend against terrorist threats. Each adopted a different approach. George W. Bush oversaw the inauguration of the Predator and was the first president to use remote warfare in high-value targeting, or targeted killing, operations. According to reporting, he also approved the first Predator strikes outside areas of declared hostilities, for example, in Yemen, Pakistan, and Somalia. Barack Obama increased the number of remote warfare attacks considerably. But he also took a personal role in identifying the terror leaders who would be killed. Famously, at a recurring Tuesday meeting that came to be known as Terror Tuesdays, President Obama and his national security team met to nominate targets. Ultimately, the president would decide which names should be added to the list. Donald Trump took yet another approach, delegating authority, even for strikes outside areas of declared hostilities, to military commanders. At the time of writing, Joe Biden's administration has suspended Trump-era policies but has not yet instituted its own.[5] However, after the US withdrawal from Afghanistan, Biden and members of his national security team have repeatedly referred to the US ability to conduct "over-the-horizon" counterterrorism operations. As early as July 2021, US Central Command had established an "over-the-horizon strike cell" to identify, track, and disrupt attacks by ISIS or al Qaeda forces.[6] Several commentators have suggested that this capability refers at least in part to continued Reaper operations in Afghanistan.[7]

Just because presidents use remote warfare doesn't necessarily mean that it's moral. But remarkably, these four presidents from across the political spectrum have continued using these kinds of tactics to disrupt and degrade terrorist networks around the globe. Whatever our initial beliefs about the morality of remote warfare, there seems to be a space for the justified use of these tools in combating transnational terror organizations.

Even as the United States continues to employ remote weapons in targeted-killing operations, access to these remote tools is increasing. For much of the last two decades, the ability to attack enemy targets with remotely piloted aircraft belonged only to a few states: the United States, the United Kingdom, and Israel. We are approaching an inflection point, though, in what many have called *drone proliferation*. States other than the United States and Israel—most notably China—have begun to export these systems. By 2019, some three dozen states around the world had either developed or imported armed, remotely controlled aircraft.[8] And those are just state actors. The use of remote weapons, even if less sophisticated ones, is increasing among nonstate actors too. In 2019, Houthi rebels in Yemen claimed responsibility for a remote attack on an Aramco oil facility in Saudi Arabia. Just a few years before that, ISIS fighters used quadcopters to drop grenades on US and allied soldiers—ending the sixty-five-year streak in which US air superiority prevented adversaries from attacking American soldiers from the air.[9]

On top of all that, we are approaching an important historical moment in which leading military powers will turn to artificial intelligence (AI) to control their weapons systems. In 2017, Russian president Vladimir Putin said that "artificial intelligence is the future, not only for Russia, but for all of humankind." Hinting at military power, Putin added that "whoever becomes the leader in [the AI] sphere will become the ruler of the world."[10] The People's Republic of China has similarly been pursuing military AI technology. In a policy document published in 2017 and titled "New Generation Artificial Intelligence Development Plan," the Chinese government outlined its goals to be a world leader in some AI applications by 2025 and to become a world leader in AI by 2030—enabling "leapfrog developments" in military capabilities.[11] The United States is also pursuing military AI and anticipates applying it in "operations, training, sustainment, force protection, recruiting, healthcare, and many others."[12]

It would be a mistake to conflate the categories of remote weapons and autonomous, or AI-enabled, weapons. Autonomous weapons are systems that can perform a series of tasks without any human intervention. AI is a general technology that enables machines to make predictions based on historical training data. The remotely piloted aircraft with which I am concerned in this book, the Predator and Reaper, are neither AI-enabled nor autonomous. But as the world's leading military powers pursue AI-enabled weapons systems, elements of remote warfare will become more common. AI will make distributed warfare the rule rather than the exception. As a result, the lessons we learn about the morality of remote warfare will become increasingly important as this new era of AI-enabled warfare unfolds.[13]

Remote warfare is changing—indeed has already changed—the character of war. But these changes are too layered and complex to be captured in simple questions of pro-drone or anti-drone or questions about whether remote warfare is morally right or wrong. The morality of war is complicated, and so is the morality of remote war. My purpose in this book is to take a more thorough look at the ethical questions raised by remote warfare. I focus on the human pilot and crew and take into account their human judgment, human virtue, and, at times, their human shortcomings. That the humanity of the crew is relevant to the ethics of the operations seems an obvious point, yet the role of the crew is often overlooked and almost always misunderstood in the ongoing controversy surrounding remote warfare. Crews are sometimes portrayed as automata who have no influence over the operations in which they participate. Or they are portrayed as gamers who cannot empathize with those on the other end of their cameras and missiles. Still others see the crews as psychologically tormented by their sometimes lethal work. In this book, I hope to offer a different, and more accurate, view of US Air Force remote warfare crews: professional airmen who do important work in the service of their country.

## THE FIRST SHOT

On October 6, 2001, the opening night of the US-led coalition's war in Afghanistan, Air Force Captain Scott Swanson squeezed a trigger and his Predator aircraft released a Hellfire missile that killed members of Mullah Muhammad Omar's security team and destroyed one of his vehicles. The missile strike was in Kandahar, Afghanistan. Swanson sat in McLean, Virginia.[14]

That event—the first strike conducted by a Predator crew half a world away—inaugurated a new way of warfare that has become emblematic of Western conflicts in the early twenty-first century. But putting a finger on what precisely was novel about that strike is not as easy as it sounds. For example, 2001 was not the first time the US military had the capability to kill from half a world away. Nuclear missile officers in US-based silos had that capability as early as 1958.[15] Nor was 2001 the first time the United States could use conventional weapons to strike targets from well outside the target area. In the 1991 Gulf War, US Navy submarines and surface ships in the Persian Gulf launched Tomahawk land attack missiles hundreds of miles into Iraq.[16] The year 2001 was not even the first time the United States attempted to strike an enemy leadership target from outside the theater of operations. As we will see, just three years before the war in Afghanistan began, the United States tried to kill Osama bin Laden with cruise missiles launched from the Arabian

Sea into Afghanistan.[17] If Swanson's shot inaugurated a change in the character of war, what precisely was that change?

Numerous commentators have attempted to answer this question. Grégoire Chamayou, Christian Enemark, John Kaag, and Sarah Kreps agree on a few central concerns. One is that the armed, remotely piloted aircraft that debuted in 2001 enabled powerful states to conduct "riskless" warfare.[18] These states rely on technology to impose impassable distance between their own forces and those of the enemy. The result is a relationship according to which combatants on one side impose risk on their enemies but face no risk themselves. The enemy faces risk but cannot impose it on the other side. According to some arguments, whatever this relationship might be, it cannot constitute *real* war. As some have argued, real war as it has existed since Hector and Achilles consists of warriors on one side fighting warriors on the other. Each poses a threat to the other, and each faces risk. Remote warfare defies this model. And so, some have argued, it cannot be real war. And, crucially, if it cannot be war, then it cannot be morally justified by the principles of just war theory or legally justified under the laws of war.

This argument has some appeal. Though Swanson remains relatively untouchable, this new weapons system gives him the capability, as US warfighting doctrine puts it, to "find, fix, and finish" enemy targets.[19] The finding—the hunt—is inherent in high-value targeting. And we might initially associate remote weapons systems like the Predator exclusively with high-value targeting operations. Creech Air Force Base outside Las Vegas is home to the US Air Force's 432d Wing, the unit responsible for the majority of air force Reaper (and formerly Predator) operations.[20] The wing members proudly call themselves "the Hunters," and the base has long greeted its entrants with a sign that says Home of the Hunters. And so the argument that this new way of warfare is characterized by the hunt has some intuitive appeal.

This twofold claim that remote warfare is defined by the hunt and that the hunt cannot be real warfare is important. If it is right, it removes remote violence from the domain of just war theory. According to many just war theorists, in the context of war, a combatant who takes the life of an enemy can justify these actions by saying merely that the person killed was an enemy combatant. That is enough. When one political community goes to war with another political community, it directs its military members to seek out and kill the enemy group's military members. But if remote warfare is not real war and therefore falls outside the scope of just war theory, then the bar for the justified use of lethal force is much higher. With this common approach to just war theory, the justification that the person killed was an enemy combatant applies only under conditions of war.[21] If remote violence is, by its nature, something other than war, then the most common justification for killing in war—that the person killed was an enemy combatant—evaporates. Therefore, according to this argument, whether remote warfare can count as real war determines whether it can be morally justified. The study of the ethics of remote warfare must address, and perhaps even begin with, an investigation of these more fundamental questions: What is remote warfare? How does it relate to what we often think of as traditional warfare? And to what principles should we appeal to evaluate its morality? These are the questions that I attempt to answer in this book.

## MYTHOLOGIES

The misgivings many people have about remote warfare and whether it is real warfare are understandable. This technology has challenged long-held conceptions about what terms such as *war*, *warrior*, and even *pilot* ought to mean. When I was an academic instructor at the US Air Force Academy, I offered cadets my perspective on whether they ought to pursue a career flying traditional aircraft or flying remotely piloted aircraft. The sense in which this technology challenges our conceptions is evident even in the vocabulary we use. Students often asked about how flying drones compares with flying real airplanes. More pejoratively, some pilots of traditional air force aircraft speak of the distinction between remote pilots and real pilots. All things considered, this distinction between real and something less than real is imprecise and unhelpful. These aircraft are real airplanes, after all. But the vocabulary does point us to the challenge these aircraft pose to our existing concepts.

In the spring of 2008, I was in US Air Force pilot training, still months from graduation. My class was told that ours would be the first undergraduate pilot training class to receive what was then called an unmanned aerial vehicle assignment. One of the five of us would be directed to fly the Predator. We would not be asked to volunteer. I do not exaggerate when I say that we were devastated. Having worked as hard and as long as we had to become pilots, the threat of being relegated to this other role seemed tragic. As the months passed, we were eventually told that the air force was not yet ready for its first group of Predator pilots direct from pilot training and that the burden would fall to a later class. My concern then was not directly with the ethics of remote warfare but with my lifelong dream of being a pilot. It was a future for which I had longed and a professional identity to which I (perhaps unjustifiably) felt entitled.

My own exaggerated conception of the distinction between the so-called real pilot and the remote pilot parallels the ethical concerns over the distinction between so-called real war and remote war. I see now that the trouble with both arguments is that each runs the risk of comparing a caricature with a mythology. My pilot training class compared a caricature of what it meant to be a Predator pilot with the mythology of the traditional pilot. Likewise, in the ethics debate, we risk comparing a caricature of remote warfare with the mythology of traditional war.

I had been taken in by the mythology of the aircraft pilot—the master of the machine, pioneer of the wild blue, and distant observer of the world below. From the very beginning of combat aviation in the First World War, combat pilots were compared to the mounted nobility of the Middle Ages. But the knights-of-the-air portrayal that these nascent aviators received was, even then, at least part publicity stunt intended to invigorate the people of countries that, like France and Germany especially, had already lost young men to the war in the hundreds of thousands.[22] On the surface, the Predator might seem like the single technological innovation that threatened the identity of the pilot as aerial knight. But, as we will see in later chapters, this approach compares the Predator and Reaper not to the actual history of combat aviation but to a mythology of combat aviation that, if it did exist, was short-lived.

There are, of course, clear, crucial differences between sitting in a cockpit that is contained within the aircraft and controlling the aircraft from a cockpit on the ground. Though remote crews still fly the aircraft, they do not *fly*. But how do these differences relate to critical questions about life, death, and morality in war? The most obvious difference between flying an aircraft from within and flying it from without is that the latter lacks the romance of the former. The remote pilot, unlike a more traditional counterpart, has not, in John Gillespie Magee's words, "slipped the surly bonds of Earth and danced the skies on laughter-silvered wings."

Magee's poem "High Flight" was not just required reading for me and my fellow officer trainees—we were required to memorize it. If this romantic conception is what we choose as the standard against which to compare remotely piloted aircraft, then surely the experience of flying a Predator or Reaper will fall abysmally short. We can easily deceive ourselves into thinking that combat airpower is and always has been a thing of beauty and that remotely piloted aircraft operations represent a stark departure. Similarly, we might tell ourselves that war is defined by the courage of, risk to, and potential sacrifice by combatants on the battlefield. Remote warfighters fail to measure up to this standard, and some might consider it unethical on these grounds.

The beauty we see in Magee's poetry, though, is not combat but flight. His poetry was inspired not by war but by pilot training.[23] Magee's was the romance of flight in a peaceful sky. Pilots of more traditional fighter, bomber, tanker, and cargo aircraft must likewise admit that in the midst of combat operations, there is little time for Magee's "tumbling mirth" or for "chas[ing] the shouting wind along . . . through footless halls of air." If we are to understand remote warfare crews, we will have to dispense with the mythology of the combat pilot. Likewise, if we are to understand remote warfare, we must dispense with the mythology of the traditional war.

Combat airpower grew out of observation balloons as early as the late eighteenth century and then into Orville and Wilbur Wright's heavier-than-air flying machine in the early twentieth. One of the students that the Wright brothers taught to fly—and who would go on to command all US airpower in the Second World War—was a young lieutenant named Henry "Hap" Arnold. Years later, Arnold described the transition from mere intelligence gathering to aerial violence in a 1936 book titled *This Flying Game*:

> At first flying machines were used only to collect information. . . . At first, as they passed to and fro on their journeys across the hostile lines, having no weapons, they waved a friendly greeting or made derisive signs to opposing airmen. But one day one of them carried a shotgun; the next another carried a rifle; all of them began studying how to prevent the rival from getting information or pictures, how to prevent his accomplishing his mission. The war in the air was on.[24]

The subordination of the romanticism of flight to the realities of combat began early.

A century later, combat pilots in Iraq and Afghanistan would likewise put their heavier-than-air flying machines to violent use. Commander Layne McDowell, a naval aviator whose story C. J. Chivers recounts in *The Fighters*, illustrates this subordination of the joy of flying to the demands of combat. Tortured by uncertainty from his first-ever combat weapons employment in Kosovo a decade earlier, McDowell wonders whether the family home he inadvertently bombed had been occupied. Now, on a subsequent deployment to Afghanistan, McDowell puts in sharp contrast the reality of combat airpower and the beauty of aviation. Before completing his last carrier landing of the deployment, McDowell recalls thinking, "I don't want that kind of haunting anymore. I'm glad it's over. I hope my days of flying combat are over." Before landing, he finds the tops of a flat cloud formation, high above the sea surface below. "He felt a rush of joy," Chivers describes.

> The cloud deck gave him a substitute for the ground, and a means to re-alize that old high [of ground rush]. . . . He reached the end of the cloud, climbed, turned around, dove, and surfed back. Racing above vapor just below his feet, Commander McDowell was happier than during any of his recent flights above Afghanistan. He was alive.[25]

Even as McDowell lived Gillespie's poetic dream in a $57 million warplane, he recognized its removal from combat airpower. McDowell had experienced combat airpower, and it had failed to live up to the mythology.

Here my position will undoubtedly diverge from that of many combat pilots. Magee's poetic dream about the means must be sub-ordinated to the prose of combat ends. I have never flown a Super Hornet in combat, but I have lived Magee's poetic dream—"slipped the surly bonds of earth" and "done a hundred things you have not dreamed of—wheeled and soared and swung high in the sunlit si-lence." And it was against precisely this backdrop that I volunteered to fly the Predator. When it was announced that the training pipeline for the aircraft to which I had originally been assigned would see delays of up to eighteen months, I began to ask what other options were available. The only other place the Air Force Personnel Cen-ter was willing to send me was Creech Air Force Base—the "Home of the Hunters." One of my many considerations as I mulled this possibility was that I had already experienced Magee's dream. I had lived the romance of flight. As an undergraduate pilot training stu-dent, I had flown the T-6 Texan II and the T-38 Talon in the prac-tice areas above northern Oklahoma, on low-level flights through Missouri, on instruments in the clouds, and in formation with three other aircraft—maneuvering the two-engine T-38 jet trainer with just three feet of lateral spacing between my wingtip and those of the other aircraft in the formation. I could check it off the list, and I was then ready to get to work. The view I held then has not changed since: Real war is not about the romance of flight. It is about the work of defending one's political community.

In the chapters that follow, I first show how remote warfare can fit into our understanding of war more broadly. But this is only one part of the challenge. Understanding the relationship between remote war and traditional war does not cause all the ethical questions to dissipate. Instead, it brings the most significant ethical ones into sharper focus.

## SCOPE

While my operational background has left me with biases, I am not inescapably pro-drone. I am, however, unequivocally pro-airmen, and I remain convinced that the military professionals who carry out these missions are rarely discussed and often misunderstood. I hope to offer a better understanding of the work they do in remotely piloted aircraft operations. By no means am I suggesting that remotely piloted aircraft crews never err. But in the absence of evidence to the contrary, they ought to be given the same presumption of professionalism and integrity that is often given to their more traditional military counterparts.

My perspective is a counterweight to those who write about remote warfare as outsiders, with a perspective that, no doubt, comes with its own set of biases. My biases, though real and unavoidable, are more asset than liability. In addition to operational flying training and experience, I have had the privilege of formally studying moral philosophy and, more specifically, how moral philosophy bears on the military and on war. In this book, I offer something you are unlikely to find elsewhere: careful analysis of the ethics of remote warfare written from the perspective of one who has both studied military ethics and participated in remote warfare.

⊕

The ethical concerns in the remote warfare literature can be divided into two broad categories: one at the level of national foreign or military policy and the other at the individual unit or warfighter level. For simplicity, I refer to them as strategic-level and tactical-level concerns.[26]

Strategic-level concerns focus on the policymakers who employ remote warfare crews as a foreign policy tool. Many observers have argued, for example, that remote warfare enables policymakers to resort to military force without the political risk associated with service members being killed in action. On this view, remote warfare is morally problematic because it encourages policymakers to use force even when they have strong moral reasons not to.[27] Another concern is that remote warfare has enabled such powerful states as the United States to employ military force outside areas of declared hostilities, with relatively little political resistance either domestically or internationally. One possible result is that while al Qaeda fighters in Afghanistan and ISIS fighters in Iraq and Syria are, or have been, lawful combatants, the actions of terrorist organizations outside areas of declared hostilities (Yemen, Somalia, Libya, etc.) are not so easily categorized.[28] A third concern in light of the rapid proliferation of these weapons systems is whether the policies established by states like the United States, the United Kingdom, and Israel have appropriately included ethical and legal norms to which other states ought to aspire.[29]

These are important—and difficult—questions. Decisions at the strategic level have weighty moral consequences. If a pilot makes a mistake about releasing a weapon or if a ground force commander errs in identifying an enemy formation, noncombatants might be unjustly killed. These mistakes are tragic, to be sure, but when a head of state errs in the decision to wage war, the result is death, destruction, and unjust harm on a massive scale. The strategic-level questions about the ethics of remote warfare are especially important because of this massive scale. If, as many have argued, the ability to conduct remote warfare emboldens leaders to engage in more wars more often, remote warfare really does have important ethical implications.

Others have asked questions about remote warfare from the opposite viewpoint. What if this sort of warfare makes powerful states more likely to fulfill their moral obligations to defend victims against unjust aggression?[30] We might think of cases such as the 1994 Rwandan genocide. Many people believe that the United States, or some other Western state, ought to have sent its military to prevent the Hutu from murdering hundreds of thousands of Tutsi victims. The United States was observing events in Rwanda from a distance and was on the heels of the Black Hawk shootdown in Mogadishu the year before. There, after Somali militia fighters brought down two helicopters and killed eighteen US service members, Americans watched in horror the video of US soldiers' bodies being dragged through the streets. In 1994, the United States lacked the political will to commit US forces to support non-US citizens in another faraway place. But what if President Bill Clinton had had the same remote warfare capability that Presidents Bush, Obama, Trump, and Biden have had? Would the prospect of a low-risk US military intervention have spurred the administration to commit US remote warfare to end the Rwandan genocide? It's impossible to know. And while we are imagining historical counterfactuals, we must also ask whether remote warfare would have been successful in halting the genocide. I have my doubts.

At the tactical level, the moral questions function in a different way. By focusing on warfare and the level of the individual warfighter or practitioner, I am better able to describe remote warfare as a category and what it means for the future of warfare. What is—and what ought to be—the warfighter's relationship to war, the enemy, or the members of the political community for whom the person fights? What is the moral significance of risk in war, the moral psychology of remote killing, or the remote warfighters' ability to cultivate martial virtue?

My intent is to draw conclusions about remote warfare through a focus on the US military's employment of remotely piloted aircraft and address specific questions of morality, warfare, and risk. As the technological and operational leader in remote warfare, the United States is setting the precedent and establishing the norms to which other states will eventually be held. Many Western states (e.g., France, Italy, the Netherlands, Australia), though investing in remote warfare, presently lack extensive experience employing it in combat. Meanwhile, some non-Western states (e.g., China, the United Arab Emirates, Saudi Arabia, Iran), though rapidly gaining experience, are not forthcoming with operational insights. Although many have called for increased transparency into US remote warfare operations, the US military remains, in some respects, transparent enough; we can paint at least an imperfect picture of its remote warfare operations using solely publicly available information.[31]

I focus on US military employment and refrain almost entirely from references to the Central Intelligence Agency (CIA) or other US government agencies. There is far more publicly available information about US military operations than about any CIA operations.

Finally, in this book, I focus on remotely piloted aircraft. There are, indeed, other remotely controlled weapons systems—land vehicles, submarines, and potentially forthcoming space weapons. But the systems with which I am most familiar, those that have dominated the literature, and those that in fact seem to have taken a permanent place in the public consciousness, are aircraft. In the end, though, this book is about the ethical implications of distance in war. So, while submarine operations and land warfare differ a great deal from air operations, the conclusions I draw in this book should be relevant to a wide array of remote warfare applications. And as the world's leading militaries continue to invest in robotics, AI, and space systems, these moral questions about distance in war will become more relevant, not less.

The remotely piloted aircraft category is broad, complex, and contentious. I generally have in mind the MQ-1 Predator, which was retired from the US Air Force in 2018, and the MQ-9 Reaper that replaced it. These aircraft are used widely enough to be important test cases and well known enough to be referred to by name. In May 2010, for example, President Obama made a joke in which he threatened the family pop trio the Jonas Brothers if they were to get any ideas about the president's daughters. "I have two words for you," Obama said, "Predator drones." The joke drew criticism on the grounds that the president was making light of such a weighty moral issue as targeted killing. But more to the point here, the president of the United States mentioned Predator drones by name.[32] Nearly a decade before, President Bush also referred to this airplane by name in a speech at the Citadel: "Before the war, the Predator had skeptics. . . . Now it is clear the military does not have enough unmanned vehicles."[33]

Presidential references to a particular weapons system are exceedingly rare. After the September 11, 2001, attacks, for example, Vice President Dick Cheney distilled the unique power of the executive branch by saying "We've got the helicopters," but he didn't say which ones. In Winston Churchill's legendary 1940 speech, in which he praised the Royal Air Force airmen who won the Battle of Britain, he mentioned fighter aircraft, bomber squadrons and machines but never mentioned Spitfires and Hurricanes. The language of commanders in chief is the language of strategy. US presidents often speak of air and naval forces, of American bombers, and of American ships, but rarely of specific weapons.[34] And yet, these two US presidents made reference to this one airplane by name—and everyone knew what they meant.

## THE PLAN

In Chapter 1, "The Predator Paradigm," I look closely at one weapons system in particular, the MQ-1 Predator. In many ways, the Predator defined the genre. It was the first remotely piloted combat aircraft to achieve sufficient success in the battlespace to secure these kinds of aircraft as a fixed category in modern warfare. If we fail to grasp what made the Predator significant in military history, we will fail to understand the whole category of remotely piloted combat aircraft.

As theorists attempt to make sense of the last twenty years of remotely piloted aircraft operations, one common claim is that armed aircraft such as the Predator and Reaper have, for the first time in the multi-thousand-year history of warfare, enabled war without risk. This assertion warrants closer attention. In Chapter 2, "Riskless Warfare?," I investigate numerous claims that war must be relevantly like a duel. Many have argued that the riskless nature of remote warfare implies that whatever this remote violence might be, it cannot be war. I argue that not only is war not necessarily like a duel but, in fact, the duel is a poor proxy for understanding the morality of war.

In Chapter 3, "The Morality and Psychology of Remote Warfare," I tackle questions of moral psychology that have been formative in the ethics debate. I show that the two broad narratives that have arisen—that remote warfare crews have a PlayStation mentality and that many of them suffer from post-traumatic stress disorder (PTSD)—both lack nuance and inadequately describe the psychology of remote killing. The truth is more complicated.

One reason remote warfare raises ethical concerns is that it seems to deviate from the traditional model of warfare in which a combatant fights for brothers and sisters in arms. But this is not obviously true for the remote warfighter. In Chapter 4, "Good Guys and Bad Guys," I ask whether remote warfare can be morally justified, and if so, on what grounds. For millennia, warriors have described battle as a fight "for the person next to you." But can this be true of a Reaper crew half a world away?

One often-overlooked aspect of contemporary remote warfare is how much human judgment crew members can apply to a dynamic battlespace. In Chapter 5, "Human Judgment and Remote Warfare," I show that, just as remote warfare shrinks the psychological distance between human warfighters and the effects of their work, remote weapons allow their crews to impose human judgment as though they were much closer. If psychological distance is about the war's effect on the crews, human judgment is about the crews' effect on the war.

In Chapter 6, "It's 'Hard Work to Be Excellent,'" I respond to claims that these contemporary warfighters cannot cultivate the traditional martial virtues. In the absence of physical risk to self, some scholars have asked, can remote warfighters cultivate the long-standing martial virtues of courage, loyalty, and honor? And if they cannot, can they accurately be called warriors? In this chapter, I argue that the concern about the martial virtues misinterprets the relationship between the warrior and the martial virtues. Rather than conceiving of the martial virtues as necessary for admission to the class of warrior, we should instead consider them instrumentally valuable as the military member pursues just aims. Which character virtues are instrumentally valuable as remote warfighters attempt to achieve just military aims?

In Chapter 7, "What Comes Next?," I look to the future. The age of autonomous weapons—or AI-enabled weapons, machine learning weapons, or whatever else they might be called—is fast approaching. We can draw some important lines between remote weapons and autonomous ones, but these connections are few. In this final chapter, I highlight the themes that have arisen throughout the book—moral justification, virtue, and human judgment—to show that even though the Predator and Reaper are controlled remotely, they depend heavily on human input. AI-enabled weapons are quite a different category from remote weapons, and many of the ethical questions they raise are different too.

# 1

# THE PREDATOR PARADIGM

IT IS DIFFICULT TO IMAGINE ALL THE WAYS THE WORLD HAS changed as a result of the attacks of September 11, 2001. As the child of a military member, I felt some of the changes in the US military's security posture firsthand. I grew up on a small air force base in Massachusetts. Before 9/11, I could walk right off the base and into Minute Man National Historical Park and to the small edifice commemorating the spot where British soldiers captured Paul Revere on his midnight ride in 1775. After 9/11, though, we became an air force at war, and, where previously there had been no barrier between the military families and the outside world, there is to this day a chain-link fence and razor wire.

The changes the US military began to undergo in the fall of 2001 were not just revisions in security posture. The military had to adapt to a new set of enemies, to new adversary tactics, and to fighting in a new theater of operations. But another change often goes unnoticed in the US transition from airpower operations of the 1990s to the post-9/11 wars—and it is central to an understanding of the Predator and Reaper. For decades, the US Air Force had learned to target various elements of a sophisticated military's warfighting apparatus: heavy weapons, integrated air defense systems, headquarters buildings, and rocket launchers—the stuff of modern military capability. In the post-9/11 wars, though, the air force and especially the crews of Predator and, later, Reaper aircraft would be asked to strike a new kind of target: people. In previous operations, there had certainly been people in the heavy weapons vehicles, the integrated air defense sites, and the headquarters buildings, but success or failure in those operations did not depend on killing the people. Success lay in destroying the systems. Targeting terror cell leaders, though, was different. Success or failure in a given mission depended on killing, or failing to kill, the person under the crosshairs.

This shift, the result of geopolitical realities, affected the air force on the whole, not just its remote warfare operations. The Predator came into existence alongside the US "global war on terror," and as a result, targeted killing came to shape much of the Predator's adolescence. This shift from targeting things to targeting people took the air force by surprise and would cause reverberations in technology, tactics, ethics, and psychology that are still felt to this day. But to understand this shift, we have to go back to the 1990s and to the development of the MQ-1 Predator.

I have an old photo of myself standing in front of a Predator, arms folded, sleeves rolled up to just above the elbow, woodland green battle dress uniform—a uniform that has now been replaced twice over. Behind me in the cavernous gray hangar is an unarmed RQ-1 Predator, its fuselage distended at the front, and its tail fins, or stabilators, pointing characteristically downward in the shape of an inverted $V$.[1] The picture, from 2003, depicts both a uniform and an airplane that have been retired. That summer, the US Air Force sent about forty officer trainees to Nellis Air Force Base,

Nevada—Home of the Fighter Pilot—for six weeks. We were offered an immersive experience across a range of air force specialties.
We rode in fighter aircraft, tankers, and helicopters. We went on
simulated aircrew rescue missions on Lake Mead with para-rescue
airmen. And we fired weapons from across the security forces' arse-
nal into the Nevada desert.

One early morning, the temperature still bearable before sunrise,
a dozen of us piled into a blue fifteen-passenger van and made the
drive up US Route 95 toward Indian Springs Air Force Auxiliary
Field. In the final decades of the last century, Indian Springs served
two functions. It was an emergency airfield for aircraft participating in large force exercises at Nellis, and it was the primary practice
airfield for the US Air Force's demonstration squadron, the Thunderbirds. In fact, Indian Springs had been the site of the most tragic
event in the Thunderbirds' history. While training for the 1982 air-
show season, flight lead Major Norm Lowry led four of the Thunderbird T-38 aircraft in a maneuver called the *line abreast loop*. All
four aircraft were in position and on altitude at the top of the loop,
some seven thousand feet above the Indian Springs desert. As Lowry
began to accelerate toward the ground and pull his aircraft's nose
toward the horizon, the three other pilots, Willie Mays, Joseph Peterson, and Mark Melancon, did just as they were supposed to do—
they kept their eyes fixed on their flight lead and based their position
and altitude on Lowry's. Unknown to anyone on the team, however,
Lowry's horizontal stabilizer had suffered a mechanical failure somewhere near the top of the loop. With the stabilizer stuck in the wrong
position, Lowry was unable to pull the nose of the aircraft through
the loop aggressively enough. As Mays, Peterson, and Melancon remained in position off the lead jet, Lowry continued to try to pull
the nose through the loop. The four-ship formation crashed into the
Nevada desert at nearly five hundred miles per hour.[2]

Following that devastating accident, Congress threatened to end the
Thunderbirds forever. General Wilbur "Bill" Creech, then commander
of Tactical Air Command, and former Thunderbird pilot, waged a

successful campaign not just to save the Thunderbirds but to replace
the T-38s that had crashed at Indian Springs with the F-16 Fighting
Falcon ("Viper"), at that time the air force's newest frontline fighter. He
was successful, and the Thunderbird demonstration squadron continues to fly the F-16 to this day.[3] Creech died on August 26, 2003—just
days after my visit to Indian Springs.[4] Two years later, Indian Springs
Air Force Auxiliary Field would be renamed Creech Air Force Base.

On that summer day in 2003, we trainees went to Indian Springs
to see the Predator. This visit was the first opportunity I had to try to
wrap my head around the distance relationships involved in remote
flight. At one point, two of us stood in the cockpit behind the pilot
and sensor operator and looked at the black-and-white infrared image on the screen. We saw a modified steel shipping container whose
image slowly rotated on the screen as the aircraft flew past. Through
the lens of a camera mounted below our airplane, we were looking at
our own cockpit. At one point, the crew had us take turns walking
to the back of the cockpit, opening the door, and looking for the aircraft in the sky. I watched the image on the screen as the trailer door
opened. The other cadet stood in the doorway. He hadn't yet found
the airplane fifteen thousand feet above the desert, but he waved in
its general direction. We watched him wave in infrared on the screen.
It was an odd thing to be looking at ourselves through the Predator's
sensor ball.

After that, our hosts took us to the hangar to see the airplane up
close. I asked a friend to take the same picture of me that I had taken
in front of the F-22, the F-15E, the HH-60, and the Thunderbirds:
arms folded, sleeves rolled up to the elbow. I left that day thinking
that what we had seen was interesting—it was noteworthy. But it
wasn't for me. A few short years later, I would be an instructor pilot
in that same squadron. I would fly from that same cockpit.

The Predator is the remotely piloted aircraft that defined the genre.
Even then, as early as 2003, it was already changing the character
of aerial warfare in the twenty-first century. And this aircraft has
been at the center of the ethical debates about remote warfare. As one

aeronautics professor put it, "Whichever side of the debate you're on, Predator has been a focal point."[5]

Beyond its symbolic role, the Predator has played an important historical role in the development of twenty-first-century warfare. The transition from the era of remotely controlled aircraft that preceded the Predator to the Predator paradigm was sudden, and its implications far-reaching. Remotely piloted aircraft development in the last century yielded some limited success. Those advances did demonstrate new technological capabilities and showed how remote weapons systems might enable new ways of operating. But these developments failed to break through into military doctrine and standard operating procedures. They attracted little attention; few senior military leaders recognized their potential. It seemed inconceivable that these airplanes would have lasting effects on the character of war.[6]

That observation no longer holds. The Predator created a new category of weapons system that has indeed provided serious combat capability. Even though the Predator has been central to the study of military remotely piloted aircraft for so many years, this aircraft, which one military researcher called an "odd bird," is still widely misunderstood.[7] By looking closely at the Predator and its historical development, I provide some plausible reasons for the widespread misconceptions. And then by understanding the Predator in the context of the operational problem it was designed to solve, we will have a better sense of the drastic shift the air force underwent after September 11, 2001—the shift from targeting things to targeting people.

What the Predator is and does and what it means for the character of war look very different from the outside and from the inside. To those on the outside—that is, to those who have never watched Predator operations from inside the cockpit—there are powerful reasons to see the Predator defying established categories. Those who fly it, however, see its continuity with more traditional combat aircraft. This perceptual difference has had a significant and negative effect on how the Predator is understood in academic, policy, and even artistic spheres.

Two strong assumptions about the Predator have emerged. The first is that it is the epitome of *push-button warfare*.[8] That is, it represents the ultimate in technologized warfare to the degree that perhaps we ought not to consider this kind of violence warfare at all. As we will see, however, although the Predator was new and different, it was also the next logical step on the air force's decades-long path to become a techno-social force.

To perceive the Predator as merely a technological system is to misunderstand the role of the human crew. Since its inception, the Predator has been a techno-social system, inclusive of both novel technology and human operators as independently necessary and jointly sufficient for Predator operations. In a techno-social system, the human operator surely has an effect on how the technology is employed. At the same time, the technology can come to shape how the humans see themselves and their role.

A second widely held assumption is that the Predator appeared on the scene as a hunter-killer weapons system that first hunts and ultimately targets individual people.[9] The Predator is often understood to be a weapon tailor-made for high-value targeting—for targeted killing. While this statement isn't untrue, it's not the whole truth. In fact, as I'll show in this chapter, the adoption of the Predator as a weapon for high-value targeting was a mere accident of history. And, just a few months after its first use in targeted killing, a Predator crew supported US ground forces engaged in a deadly firefight with the Taliban. As we will see, almost from the beginning, the Predator was a close air support aircraft as well as a high-value targeting aircraft.

## THE PERCEPTION OF AUTONOMY

The Predator has often been described as a type of autonomous or robotic weapon. It is easy to see how these misunderstandings came about—after all, it was a brand-new kind of weapons system—a whole new kind of military aircraft. It takes time to understand where a new technology fits alongside other technological categories.

A few distinctive features about how this airplane looks can naturally cause viewers to ascribe to it a kind of autonomy that it just does not possess. In fact, as we will see, the human pilot has been central to Predator operations from the very beginning.

From its debut as an armed aircraft in 2001 to the first Reaper sorties in 2007, the Predator stood in a category by itself. The development of a brand-new, unique aircraft group is a rare phenomenon in military aviation. Ordinarily, new aircraft fall into long-standing and well-established operational categories. For instance, though the F-22 Raptor incorporated novel technology and included new features, it was immediately recognizable as an air superiority fighter. And if we want to know what it means to be a fifth-generation air superiority fighter, we can refer to antecedents such as the fourth-generation air superiority fighter—the F-15 Eagle. A Raptor exists to do what an Eagle does, but better. These robust operational categories give us a conceptual advantage when we are faced with a new weapons system that falls within an old category. We know what we mean by *strategic airlift*; *lightweight, multirole fighter*; and *long-range bomber*. Many of the most high-tech and high-cost twenty-first-century combat aircraft acquisitions fit neatly into these prepackaged conceptual categories.

This was not the case with the Predator. Even a nuanced look at Predator's closest antecedents will be of little help. The armed Predator that debuted in 2001 represented a union of remotely controlled aircraft, a multispectral targeting system (the collection of cameras and lasers that constitute the *targeting pod*), and the Hellfire missile. The Hellfire missile came to the Predator via army and marine corps attack helicopters, and the multispectral targeting system through air force special operations AC-130 gunships. If we want to understand what the Predator is and what it means for the character of war, it is not enough simply to look at twentieth-century remotely piloted aircraft, nor is it enough to say merely that the Predator is a derivative of drones, helicopters, and gunships. We need a new conceptual category.

A century of combat aircraft development generally followed a trend: new aircraft were similar to previous aircraft, just bigger and faster and capable of going higher and farther. Periodically, however, a technological development takes place outside these trend lines. The union of a remotely piloted aircraft, a multispectral targeting system, and the Hellfire missile amounted to a new operational category, just as the Wright Flyer, the submarines of the US Civil War, and the first intercontinental ballistic missiles (ICBMs) represented new categories of military platforms. Each innovation challenged standing conceptions and demanded that military operators, strategists, and policymakers modify old ways of thinking. As the first in a new category of weapons system, the Predator was an operational orphan.

Not only did the Predator defy old categories of aircraft, but its aesthetic is inhumane and sends a barrage of signals. The design is utilitarian—form follows function in this case. While these signals were not intentional on the part of designers, they are strong and clear. These visual cues—the bulbous nose, among other things— signal not just physical distance between human and machine but the absence of humanity altogether.

In the century since the Wright brothers flew at Kitty Hawk, North Carolina, aircraft have come in myriad sizes and configurations: monoplanes, biplanes, triplanes, pusher props, jet engines, multi-engines, helicopters, flying wing, swept wing, swing wing— the list goes on. But in nearly every case, there is at least one pilot on board. And, at least since the period between the first and second world wars, when aircraft could achieve speeds that required windscreens, the pilot looks through glass at the front of the airplane. Throughout this history and at airports today, we are exposed to thousands of images of aircraft. Almost every aircraft includes a place for the pilot and glass through which the pilot can see and be seen.

The canopy glass or flight deck windscreen has become the visual representation of the human who employs the system. In an important sense, the canopy glass is to the aircraft as eyes are to the person. It is the window through which we can look into the machine to see the human person or persons responsible for the machine. For a hundred years, it was the means by which the pilot—as surrogate for the aircraft—saw the world and the medium through which the outside could see the pilot. The ancient proverb tells us that the eyes are the windows to the soul. In much the same way, the canopy glass or windscreen is the medium through which we perceive the human being inside the machine—the canopy glass is literally the window to the soul onboard.

Our ability or inability to see the human operator has an important effect on our perception of the machine. When I stand in the crosswalk, unsure whether the approaching driver will yield so I can cross, I look through the windshield to see the driver. It is the car that poses the threat, but it is under the driver's control. When I was living in England, it took me some time to condition myself not just to drive on the other side of the road but also to look for the other driver on the opposite side of the car. For the first few weeks, I looked expectantly at an oncoming car's front left side often to see only the disconcerting image of an empty seat (or a dog!). As we attempt to understand the traffic and predict what drivers might do, we need to look through the windshield and make eye contact with the driver.

We are now entering the age of self-driving cars, some of which are already on the road. Like the Predator, these cars do not rely on the windshield glass for situational awareness about the world around them. Instead, they use cameras mounted on the outside of the machine. Imagine standing in the crosswalk as a vehicle approaches. It is not autonomous but is instead driven by a human driver who relies on cameras and video feeds rather than the windshield to see outside. Finally, imagine that the windshield glass—for which the driver has no need because of the cameras—is replaced with the dull gray opacity of a laminate skin. Even if we knew full well that the driver could see by using the cameras, I would be unnerved; wouldn't you? As the opaque car begins to slow, and you cannot yet be sure whether it will come to a complete stop, would you cross in front of it?

The transparent medium through which I have, my entire life, viewed the human in control of the vehicle has been replaced. Where I am conditioned, through hundreds of thousands of previous instances, to see a human operator, I see only dull composite gray. Our perception of this techno-social system that consists of both car and driver is immediately skewed. Is it a techno-social system or just a technological one? I know there is a driver, but is this person really in control? Can I trust this new thing that manifests none of the human characteristics of its human operator? What visual cues are available to me on which to make this kind of judgment?

The difficulties from the car scenario are significant in the Predator's case. How are we to perceive the humanity of the human Predator crew if, where we are conditioned to see the pilot through the canopy glass, we are given instead only dull gray? The great irony here is that unlike many other aircraft, the Predator does, in fact, have an eye. The original impetus for the Predator was not its missiles but its ability to wait and to watch—to use video imagery to find adversary targets. But the eye is mounted below the fuselage, where we are not prepared to see it. Where we ought to see the glass and a place for the pilot, we see, as one author puts it, the "bulbous nose where a cockpit would normally be."[10] Another says, "Its bulbous front . . . [gives] it a feeling of barely containable brainpower."[11] It seems a small thing, yet the absence of a visual depiction of the pilot and crew has had an outsize effect on how people perceive these airplanes.

The Predator's dull, composite, bulbous nose originally grew out of the need to pass imagery back to the United States via satellite. The original Predator, tapered at the front, looked more like a missile than a traditional airplane. But to make the airplane capable of sending imagery back to Washington, designers added to the front of the fuselage a satellite communications antenna that pointed up toward space. The antenna had to be covered by an aerodynamic radome to protect it from airflow and weather—thus, the bulbous nose. The location of the satellite dish radome is exactly where we have been conditioned to expect canopy glass and a pilot. Yet the bulbous nose gives the viewer only the opaque gray mask hiding electrical servos and bundled cables.

Evidence from psychological research suggests that humans are more likely to trust robotic systems if those systems are intentionally anthropomorphized. In one study, for example, people used a driving simulator that allowed them either to drive a normal car, to ride in a self-driving car, or to ride in a self-driving car that that had been anthropomorphized—it was given a name, a gender, and a voice. Predictably, participants self-reported greater trust in the anthropomorphized autonomous vehicle. Similar results have been found across a range of applications.[12] Humans are more likely to trust non-human systems if they perceive humanity in it—even if they know the human is fictitious. It would not stretch the imagination to suppose that the inverse might also be true. Perhaps humans will trust a system less—even a system they know is controlled by humans—if they cannot perceive human characteristics. The Predator falls into this latter category. It is a techno-social system that looks like a merely technological one.

This problem is compounded by the additional heuristics that have come to accompany discussions about autonomous weapons. Even though the Predator and the Reaper are not autonomous and are flown by a fully trained and qualified aircrew, the image of these two aircraft often serve as the front cover, the backdrop, the logo, or the banner for books, conferences, reports, websites, and just about anything else that pertains to lethal autonomous weapons. This less-than-careful application of the imagery, combined with the absence of glass where we expect to see eyes, produces impressions—even subconscious feelings—about the lack of humanity behind this weapon. Remote weapons offer important lessons that we can also apply to AI weapons—a point to which I will return in Chapter 7. But it is a mistake to think of remote weapons such as the Predator and Reaper as autonomous.

The Predator, despite its looks, is just as reliant on human inputs as is any other modern aircraft. For instance, the Predator and Reaper can fly in various autopilot modes through a series of waypoints and can maintain altitude and airspeed without continuous pilot control inputs. But these features are common to many modern military aircraft. Flight management systems on the C-17 Globemaster III and the Boeing 777, for example, are also capable of this level of (so-called) autonomy.[13] In fact, many modern aircraft now have an auto-land capability.[14] And for two decades, the Predator and Reaper have relied on the same piloting skills needed to land any other aircraft (though some say the landing is more difficult without the seat-of-the-pants feeling that traditional pilots experience). Some observers have gone as far as to suggest that the control stick and throttle present in the Predator cockpit serve no function other than as a "morale stick" to support "an illusion of hands-on control."[15] But this description is inaccurate. The control stick sends electrical signals to the electric motors that manipulate the flight control surfaces. Certainly, the control stick is sometimes bypassed, just as it is in other aircraft. When the autopilot functions are engaged, the control surfaces are managed by a computer that maintains whatever altitudes, airspeed, or roll angles the pilot has commanded. These control mechanisms are not that different in practice from those in other aircraft.

These points might seem trivial, but they help counterbalance the perceptions that the Predator's aesthetics invite. It is also helpful to see how this aircraft is similar to traditionally piloted aircraft. In fact, one reason for the Predator's ultimate operational success and its adoption by the US military was that its designer recognized the need to approach the new aircraft's operations in the same way designers and aircraft test operations have approached flying for a century.

One reason for the Predator's success as a weapons system has been the institutional insistence that the airplanes be flown by trained pilots.[16] From 1960 to 1981, the US government had invested approximately $50 billion in remotely piloted aircraft—then often called *unmanned aerial vehicles*. Yet, on average, each aircraft flew only a few dozen sorties before crashing. Abraham Karem, the Predator's designer, worked on a predecessor called the Amber and insisted that the solution would primarily depend not on technological fixes but on improved operator performance. In Karem's words, what the fledgling program needed was a "serious manned aircraft approach."[17]

Because the Predator has become the iconic remotely piloted aircraft, one might take today's ground-based cockpits for granted. In the Predator's cockpit, sometimes called a *ground control station*, the pilot and sensor operator sit side by side and observe aircraft performance and sensor inputs from the myriad displays arrayed before them.[18] Early on, however, most pilots of remotely controlled aircraft were given no first-person perspective and were instead asked to control, and even to land, the airplane from the external perspective on the ground looking up at the airplane, much as hobbyists with remote-control airplanes do. Karem made two crucial contributions. First, he put a camera in the nose so that pilots could have a view close to that of a traditional pilot from inside. This change was especially helpful in the landing phase. And second, he built a flight training program for the operators who had been conducting the Amber's flight testing. By 1988, Karem and his team had accrued 650 flight hours with only one incident—an engine failure that had been outside the operator's control.[19]

This adherence to long-standing aviation norms was revitalized in the next decade. Until 1997, the unarmed Predator was still a US Army program. Like many of the uninhabited systems that came before, the Predator was plagued by malfunctions, crashes, an absence of crucial spare parts, and poor support infrastructure.[20] Then air force chief of staff, General Ronald Fogleman, recognized that the primary reason for the army's "lousy record" with remotely piloted aircraft was that "they saw these things not as aerospace vehicles, but as just another piece of equipment, like trucks in a motor pool."[21] When the Predator program transitioned from the army to the air force in 1997, Fogleman aimed to fix the reliability issues with the same philosophy Karem had employed with the Amber—the same serious approach to safety and reliability that applied to traditionally piloted aircraft.

The US Army had already been responsible for the Hunter, a program that had resulted in numerous crashes.[22] Part of Fogleman's motivation to wrest control of the Predator—a more sophisticated system than the Hunter—was safety. "If the Army took Predator," Fogleman later said, "they would just screw it up and the program would go down the tubes."[23] The distinction wasn't mere rivalry—even if interservice competition did play a role.[24] Instead, the questions Fogleman sought to answer were basic: What kind of thing was this new system? To what operational category did it belong? Because it had no crew members onboard, perhaps it should be treated less carefully, like a vehicle in the motor pool. On the other hand, it is a multimillion-dollar aircraft. Even with no people on board, perhaps it should be managed with the same diligence and care with which other aircraft are treated.

Ultimately, Fogleman was committed to having only officers with a pilot aeronautical rating fly the Predator, because he believed their pilot training and aviation standards would improve the aircraft's reliability. "If Predator fails," Fogleman insisted, "it won't be because of our pilots."[25] As early as 1996, the air force defined a set of skills required of Predator pilots. The set mirrored the skills required for traditional pilots and included everything from knowledge of aerodynamics and meteorology to navigation and airmanship skills.[26]

Fogleman's 1996 decision to require the same level of training and evaluation for Predator pilots as for all other pilots has ramifications to the present day. When traditional aircrew members are involved in a mishap, they are required to undergo testing to ensure that they were not, for example, under the influence of drugs or alcohol at the time. The involved crew must go to the medical clinic and, as aircrew members put it, "pee and bleed"—provide urine and blood samples to the lab to test for alcohol or illicit substances. This same requirement was applied to Predator crews. Even today, when Reaper crew members crash an airplane, before an investigation has determined the cause, they proceed directly to the clinic to "pee and bleed." Fogleman's decision to hold Predator crews to the same professional standards to

which traditional pilots are held is one of the most important historical moments in the history of remotely piloted aircraft.

When I was deployed to Kandahar, Afghanistan, in 2010, I once
launched an aircraft that was out of gas. On initial takeoff, we heard
an alarm indicating that we were low on fuel. I requested an immediate landing and the airplane was fine, but it was a mystery to me how
it could have happened. I had done the walk-around of the airplane
and checked the forms just as I had always done. The maintainers are
required to fill the gas tank, indicate in the aircraft forms how much
gas they loaded, and then sign the forms. In this case, in an effort to
save time, the maintenance technician marked the forms first and
then forgot to fill the airplane with gas. When I reviewed the forms
and saw that the airplane had been loaded with gas, I went on my
way. Then I uploaded into the aircraft's software the amount of fuel
that had been written on the forms. Both the airplane and I now believed it was full of fuel. The error was a clerical one. It was the midnight shift, we were all tired, we launched several aircraft per night,
and the maintainer had simply made a mistake. But the standards are
high in aviation, and that maintenance technician was sent home. I
don't mean he was sent back to his room for the night. I mean the
contractor put him on an airplane back to the United States. He no
longer serviced our maintenance contract. Aviation is a high-stakes
field, and the decision to treat the Predator with the same diligence
and care that other aircraft received was a crucial moment in its history and one important reason for its success.

Insisting on pilot standards yielded another important element in
this techno-social system, namely, command responsibility. According to US Air Force instructions, the "Pilot in Command, regardless
of rank, is responsible for, and is the final authority for the operation
of the aircraft."[27] Operationally, just as infantry companies and cav-
alry regiments have commanders, so do aircraft. Historically, when
senior commanders send aircraft out from the airfield to conduct
operations over the horizon, those senior commanders had to dele-
gate command responsibility to someone capable of responding to

battlespace dynamics. The unpredictability of these challenges meant that someone with command authority had to accompany the crews to their targets both to maintain responsibility and to issue orders. This person was, on each individual aircraft, the pilot in command. As formations broke up, as aircraft were engaged, as pop-up threats emerged, there had to be someone who maintained responsibility for the aircraft and crew and who could issue command directives that were responsive to a rapidly changing battlespace.

Though pilots of Predator and Reaper aircraft continued according to this professional norm, there is a sense in which these pilots in command must maintain an even greater degree of responsibility than that held by their more traditional counterparts. Generally, new pilots operate on a kind of apprenticeship program. In heavy aircraft with multiple pilot positions, a new pilot begins as a copilot and must meet various training and evaluation requirements before being upgraded to the pilot in command role. In various subcommunities in the air force, this role might be called the pilot in command, the aircraft commander, or simply the A-code. In fighter aircraft, new pilots are pilots in command of their single aircraft, but those aircraft operate in formation. The new pilot, therefore, acts as an apprentice to the flight lead. Eventually, after meeting various training and evaluation requirements, the pilot can be upgraded to a two-ship flight lead, then a four-ship flight lead.

The operational demands on the air force Predator and Reaper crews have not allowed for the same apprenticeship model. On a Predator pilot's first combat mission after her check ride, she is ordinarily neither a wingman in a two-ship formation nor a copilot on a flight deck. She is the pilot in command of a single-ship combat aircraft. Its mission, its crew, and its weapons are under her command—a unique phenomenon in the US Air Force. I spoke with one Reaper pilot who had also flown C-130 airlift aircraft. He described this responsibility as "unappreciated" by the public:

> [In the C-130], I was a copilot—so I wasn't expected to make any decisions. I put the information into the . . . computer, I loaded modes

and codes [into the identification friend-or-foe system], and then I told the aircraft commander that he had a good landing. That was about the extent of it. Here [as a Predator or Reaper pilot], you're an aircraft commander immediately. And oh, by the way, the tactical situations are difficult.[28]

Another pilot with whom I spoke had flown twelve other types of aircraft before moving to the Reaper. He described the evolution of leadership with these aircraft:

In fighters, you show up as a wingman, and you're not a flight lead yet. Then as you mature and get better, you become a flight lead. But you're still under the leadership of the strike lead. By the time you're a strike lead, you've got six or seven years of experience under your belt. Here, brand-new lieutenants are pilots in command of a single-ship airplane.[29]

I don't want to belabor the point, but as we will see later, the fact that pilots are endowed with command responsibility—both for the aircraft's weapons and for its crew—will be a recurring theme that further reinforces the idea of the Predator as a techno-social system. In addition to technology, the system includes the humans who operate that technology. It cannot be understood solely with reference to the hardware. Instead, if we are to understand the Predator as a techno-social system, we must also look closely at the humans who interact with it.

## THE ROAD TO HIGH-TECH WEAPONS

From an outsider's perspective, it might look as though the Predator burst on the scene in 2001 and changed everything. The truth is more complicated. The military operational, doctrinal, and technological context shows that although the Predator might be an effect of the trend toward technologizing war, it is not the cause.

What has been called push-button warfare seems so conceptually different from the hand-to-hand combat of antiquity that perhaps we should not consider it combat at all.[30] As it turns out, however, the concept of drone operations as push-button war is much older than the Predator. As early as 1946, Hanson Baldwin published a *New York Times* article titled "The 'Drone': Portent of Push-Button War."[31] Five decades later, the Predator seemed to deliver on that prescient 1946 article. Technology, or so it seems in Predator operations, has become the medium through which Predator crews strike their targets. Some critics have been troubled by Predator operations' reliance on technology to bring about violent ends. Can it really be that this kind of violence, mediated as it is by ones and zeroes and satellite communications, should be considered combat? But as we consider airpower in its own right, we need to recognize that technology is inherent not only to remotely pilot aircraft operations but also to aircraft operations in general. Technology is intrinsic to airpower. The Predator represents not an outlier but a continuation of this long trend.

The aphorism "The Air Force . . . worship[s] at the altar of technology" has become a trope among US military professionals.[32] Behind the hyperbole, however, lies more than a kernel of truth. War in the air depends, after all, on technology.

Airpower is often described by strategists as the exploitation of the third dimension.[33] Land warfare has always taken place on a two-dimensional battlefield. Forces might be outrun, outflanked, or outgunned. But these relationships play out on a battlefield that can be represented by a two-dimensional map. The same is true of war at sea—at least until the advent of the submarine. The third dimension is the vertical—the space above the battlefield. The claim that airpower exploits the third dimension is accurate, but it is also a little misleading. Soldiers and sailors had learned to exploit the third dimension long before they learned to fly. Henry V's longbowmen at Agincourt were successful because their weapons exploited this third dimension. The longbows were an early indirect-fire weapon, and like all indirect fires, they took advantage of the vertical. Developments in gunpowder, iron, and steel enabled cannons that increased the range and effects of indirect-fire weapons and continued to exploit the third dimension. These technological developments, though, did not amount to war in and from the air.

The combat aircraft of the First World War exploited the third dimension and enabled warfighters to achieve military objectives from the air. As soldiers fought a devastating land war of inches, aircraft bypassed the trenches with ease—at least for a time. The advent of these combat aircraft altered the character of war. For the first time in military history, when faced with impenetrable enemy lines, airmen could go "over, not through."[34] This proposition grew to maturity during the Second World War, in which airpower was a crucial component of the Allied campaigns. As President Franklin D. Roosevelt famously said of Hitler's defenses, "Hitler built walls around his 'Fortress Europe' but he forgot to put a roof on it."[35] Allied airpower in that conflict sought, among other things, to exploit the third dimension to attack this perceived weakness in the Wehrmacht. None of these applications of force from the air are possible without technological or techno-social systems. Aerial warfare—the ability to go over, not through—is an inherently technological proposition. Humans can *fight* without machines, but they cannot *fly* without them.

Technology is a necessary condition for fighting on the sea surface, undersea, in the air, in space, and in cyberspace. Land war, in this regard, is the exception rather than the rule. In the ancient history of traditional war, though swords, spears, muskets, and cannon were all novel technologies in their day, there is still something reductive about the land fight. Military historian Cathal J. Nolan, for instance, describes the eighteenth-century battles in which soldiers on both sides, having exhausted their technological means, resorted to beating each other "with spent fusils used as clubs."[36] This kind of resort to nontechnical means is physically impossible in the air.

The Predator's unique contribution to the history of warfare isn't its technologically mediated operations; all airpower is mediated by technology. The claims that the Predator uniquely represents push-button warfare would imply that this aircraft relies on technology more than other airpower applications do. Or these claims would suggest that the Predator uses technology in some special way—somehow different from previous airpower applications. But these claims are also difficult to accept, given the operational evolution of the Predator. There are important—and obvious—ways in which the Predator differed from its traditionally piloted predecessors. Yet the Predator is both causally and operationally tied to the same challenges that faced US airpower operations at the end of the last century.

The Predator came about as the solution to a specific operational problem—that of mobile targets. The problem arose in the 1991 Gulf War, when US forces were unable to find and strike mobile Scud missile systems.[37] Even when US forces could find the targets, the crews took too long to target the missiles, and the Scuds would reposition. If the US military was going to strike fleeting and mobile targets like the Scuds, it would have to close what was then often called the *sensor-to-shooter loop*—the process that connects the sensor platform that identifies the target and the shooter platform that can release a weapon to strike the target.[38]

In 1993, the US Air Force faced a similar challenge in finding mobile heavy weapons systems in Bosnia. The still-nascent Predator was deployed to Bosnia from 1995 to 1997, but in its current configuration, it was of little help. The Predator crew could view video in full resolution, and the aircraft did remain over the target area for hours on end—much longer than the intermittent coverage made possible by satellites. But the crew could transmit back to Washington, at first, only snapshots and, later, poor-resolution video that was of little operational value.[39] The sensor-to-shooter loop remained a lengthy process.

In 1999, the Predator was deployed again, this time to Kosovo. General John Jumper, then commander of US Air Forces Europe, requested that the Predator program office add a targeting laser to the aircraft's camera system. Previously, the Predator crew would visually identify an enemy target—a mobile weapons system—and then verbally talk fighter or bomber aircraft onto the target. Once the fighter or bomber aircraft crew could see the target, they would then set up a bombing run to strike. But time is a precious commodity when the targets are mobile, and this process was slow. With the new targeting laser, Predator crews could now pass coordinates for the general target vicinity to the strike aircraft that would release laser-guided munitions. Those munitions would automatically track the laser provided by the Predator.[40] Called a *buddy lase*, this procedure tremendously cut down the time between target acquisition and target engagement, tightening the sensor-to-shooter loop. For the first time, the Predator demonstrated how it could support the air force's goal of striking time-sensitive targets in "single-digit minutes."[41]

In the spring of 2000—just a year before al Qaeda's September 11 attacks against the United States—the Predator was deployed to Afghanistan.[42] By this time, the Predator program office had developed something called *remote split operations*.[43] This shift from simple satellite control to remote split operations, though nuanced, is crucial to understanding the Predator and, ultimately, remote warfare.

Remote split operations is a defining feature of Predator and, later, Reaper operations. This technology allows crews to fly operational missions from the other side of the world. Imagine dozens of communications satellites orbiting the earth. Each one can "see" a certain section of the earth's surface. The original satellite communications technology the Predator employed required that the crew in the cockpit, the satellite dish, and the airplane all remain within the same satellite footprint. So the Predator could be remote-controlled, but it couldn't be controlled from the other side of the world. Remote split operations changed that. It allowed planners to put the cockpit in the United States, the satellite dish somewhere in Europe, and the airplane in Afghanistan. Remote split operations was the technological breakthrough that enabled warfare from seven thousand miles.

The most iconic modification to the Predator was the addition of the Hellfire missile in 2001. In the preceding decade, the air force had built momentum toward solving the sensor-to-shooter problem with a network of aircraft, and so there had been little motivation to arm this reconnaissance airplane. That all changed when a Predator crew observed a tall man in white robes in Tarnak Farms, Afghanistan.

Though it would not become a household name until 2001, in the waning years of the last century, US intelligence officials became increasingly concerned about al Qaeda. The most devastating of al Qaeda's attacks up to that point had been the truck bombs that detonated outside the US embassies in Kenya and Tanzania on August 7, 1998, killing 224 people and injuring about 5,000. US national security agencies soon realized that bin Laden and al Qaeda had been behind the attacks.[44] Almost exactly two years after the attacks, a Predator crew, flying southeast of Kandahar, followed "a tall, white-robed man surrounded by a security detail."[45] Analysts believed it was bin Laden, and President Clinton was willing to authorize a strike against him. In September 2000, the decision makers had a limitation: they had a sensor but no shooter.

The US Air Force built a plan to arm the Predator with a Hellfire missile, turning an intelligence aircraft designed to find Scuds into a combat aircraft trained on an al Qaeda leadership target. On September 11, 2001, while the Predator-Hellfire combination was still undergoing testing, bin Laden's al Qaeda attacked the World Trade Center buildings in New York City and the Pentagon in Washington, D.C., killing nearly three thousand. That event would have a profound influence not just on the Predator but also on the air force's approach to targeting.

Even this context—the pre-9/11 search for bin Laden and the post-9/11 war against al Qaeda—leaves us with questions about what, specifically, was new and different about the Predator. If we are to understand how the Predator changed warfare, we are going to have to look more closely at what came before.

There are several comparisons worth drawing. First, the ability to carry out intelligence, surveillance, and reconnaissance functions from the other side of the world was not, in fact, unique to the Predator. Long before its advent, satellites enabled the United States to conduct these intelligence gathering operations around the world. By the early 1970s, US satellites had become so capable that ongoing development of remotely piloted aircraft was canceled, and by 1974, the National Reconnaissance Office shifted its focus entirely to satellites.[46]

Even when we consider the addition of the Hellfire missile in 2001, the ability to strike one's foes from the other side of the globe was not entirely new either. Since the early years of the nuclear age, the United States and the Soviet Union and, later, other states have been able to threaten adversaries with ICBMs. The differences in application, however, are significant, even if obvious. Aside from the clear differences in the explosive yield of the weapons, there is the matter of latency. Because of its lengthy parabolic flight path, the ICBM's time of flight is measured in double-digit minutes. This latency nearly disappears with a Predator and its missiles, the flight time of which is measured in seconds.

The Predator offered the US two crucial capabilities, one in intelligence collection and the other in strike missions. The Predator offered a low-cost, low-risk tactical solution to intelligence gathering. More flexible than previous satellite solutions, the aircraft sidestepped the risk of a downed pilot, a problem plaguing previous high-altitude intelligence aircraft.[47] The second capability was the weapon. Once armed in 2001, in addition to the flexible, persistent intelligence gathering over the target, the system now provided a means of striking with minimal delay. For the first time, one aircraft enabled the crew to conduct all steps of the dynamic targeting process: find, fix, track, target, engage, and assess, in the words of General Fogleman, "anything that moves on the surface of the earth."[48] That ability captures the impact Predator has had on twenty-first-century war.

## KILLING PEOPLE OR BREAKING THEIR STUFF?

This shift from targeting weapons to targeting individual terrorist leaders and fighters was tectonic, and not just for the Predator crews. It came as a shock to the broader US Air Force. From our perspective now, two decades into counterterrorism and counterinsurgency operations, it is difficult to remember just how fundamental this shift was. To get a sense for how central machines have been to US Air Force culture, consider these two examples: how we determine who is a fighter ace and how we mark aerial victories over the adversary.

In the long history of land warfare, a warrior's main objective has been to kill enemy combatants. From the ancient Greek phalanx to the trenches in France and Germany, winning was often dependent on killing more of the enemy's people than the enemy could kill of one's own.[49] But warfare is different for the fighter pilot who aims not to kill the enemy pilot but to destroy the enemy's machine. The US Air Force's recognition of air-to-air victories is an indication of the centrality of the machine to fighter culture.

A pilot becomes an ace with five air-to-air kills against the enemy. Begun in the First World War as a public affairs effort, the *ace* moniker was formally adopted by the French, Germans, and Americans and informally by the British and, eventually, the world over.[50] But a subtlety in this definition of an ace often goes unnoticed. The fighter ace's five or more air-to-air kills refer not to enemy pilots but to enemy aircraft. Suppose, for example, that a US Air Force fighter pilot employs the aircraft's machine gun against an enemy fighter. The enemy aircraft is damaged beyond recovery and will ultimately crash, but before it does, the enemy pilot ejects and safely parachutes to the ground. Perhaps that enemy pilot will fly another aircraft tomorrow and continue to participate in the war. The US pilot still receives credit for one aerial victory. The idea of an air-to-air kill is agnostic as to the fate of the enemy pilot. The fighter ace is defined by five kills where *kill* is an entirely machine-centric concept.[51]

Not only does the US Air Force track victories for each pilot, but it also tracks victories for each aircraft. Independent of the pilot's record, an aircraft will have stars, enemy flags, or aircraft silhouettes painted on the side to depict the number of enemy aircraft it has shot down. These victory markings credited to the aircraft further illustrate just how machine-centric the US Air Force culture has become—right down to how we identify who the best pilots are and how we annotate victories.

For better or worse, the air force at the beginning of this century was shaped by fighter culture.[52] For an institution that worships at the altar of technology, applying a machine-centric approach to air-to-air combat to the Scud and heavy weapons hunts of the 1990s was only a minor shift. But in the post-9/11 world, the entire institution would attempt to shift focus toward finding, fixing, and finishing individual persons.[53] This change in the moral landscape of aerial warfare took the US Air Force by surprise. I will return to the implications of this shift in Chapter 3. For now, I suggest only that the shift was the result of geopolitical realities and it affected the air force on the whole—not just remote warfare operations. While the targets against which air force aircraft, including the Predator, were applied changed dramatically after 9/11, the advent of the Predator was neither cause nor effect. The two—the Predator and the US "global war on terror"—came into existence almost independently of one another and almost simultaneously. Yet, high-value targeting would come to shape the Predator's adolescence.

One recurring theme in the controversy about remote warfare is that some have understood remote warfare as a panacea for all US security concerns. It's an important concern, often unfortunately described with the adage "when all you have is a hammer, everything looks like a nail."[54] It is, however, a metaphor not well suited to the circumstances. The Predator was devised to persist over the target area for hours and provide intelligence to decision makers. When the US military made the transition to wars in which targets were not technological weapons systems but were individual combatants, the Predator was adapted to that role. But the Predator is not the US military's only tool for conducting high-value targeting operations. Similarly, targeted killing is not the only appropriate use for this tool. Since the start of the US war in Afghanistan, the missions of the Predator—and, later, the Reaper—have broadened.

Just five months after Scott Swanson's inaugural shot in Kandahar as discussed in the introduction to this book, US special operations forces would direct a Predator to employ a Hellfire missile on a hilltop 250 miles to the northeast. The Battle of Takur Ghar went badly almost from the beginning.[55] The al Qaeda fighters on that mountain were experienced, well armed, and prepared. By the end of two days of fierce fighting, seven US special operators would be killed.[56] Among them was US Air Force technical sergeant John Chapman, a combat controller attached to a US Navy SEAL team.

Chapman was originally awarded the Air Force Cross for his heroic actions under fire. Chapman had been shot and lost consciousness for a time. The SEAL team to which he was attached believed he had been killed and literally left him for dead.[57] When he awoke, he assaulted an al Qaeda bunker, killing the enemy fighters inside, and then fought on for a full hour. He ultimately left his defensible bunker position to draw enemy fire away from the inbound helicopter containing the quick reaction force dispatched to rescue Chapman and the SEAL team.[58]

In 2019, Chapman's Air Force Cross was upgraded to the Congressional Medal of Honor. Though many of Chapman's heroic and selfless actions took place after the SEAL team had left the top of the mountain and before the quick reaction force arrived, a lone Predator—this one unarmed—circled high overhead. It had no weapons with which to support Chapman, but it did record his selfless final moments. That video served as a crucial part of the air force's case for upgrading Chapman's award.[59]

This use of the Predator validated the mission for which the aircraft was originally designed—long-loiter intelligence, surveillance, and reconnaissance. Though Chapman was utterly alone on that mountaintop, a pilot and a crew some seven thousand miles away helplessly observed his every move. It was an early use of the Predator for what is now called *persistent intelligence, surveillance, and reconnaissance.* Moreover, had the entire fleet of Predator aircraft been armed, Chapman's heroic actions on Takur Ghar might have turned out quite differently.[60]

The army's quick reaction force that arrived just after Chapman was killed would call in the first-ever Predator strike in support of friendly forces in contact with the enemy. Another air force combat controller, Staff Sergeant Gabe Brown, called in air strikes against the al Qaeda fighters who were dug in and firing at the special operators. First, Brown controlled a two-ship of F-15E Strike Eagles that attempted strafing gun runs against the enemy's fortified position. Though effective, these attacks were insufficient, and the team continued to receive effective fire from the al Qaeda bunker. The second set of aircraft, a two-ship of F-16 Vipers, was armed only with unguided "dumb" bombs. These Vietnam War–era weapons had no laser or GPS guidance systems. Instead, the pilot would use onboard software to calculate the parabolic flight path of the bomb and then release the weapon at the right time, allowing gravity and the wind to determine precisely where it would fall. Brown attempted to have these aircraft "walk" the bombs up to the enemy position one at a time to make dropping on his own people less likely. In the end, Captain Nate Self, the army platoon leader, decided that the risk was too great and directed Brown to call off the attack.[61]

The unarmed Predator that had seen Chapman's final actions had returned to base and been replaced by an armed Predator, call sign "Wildfire," piloted by then Captain Stephen Jones. Brown was unaware that any Predators were armed, and it was Self who told Brown to ask the pilot about the aircraft's armament. On receiving confirmation that this Predator was armed with two Hellfire missiles, Brown directed strikes against the enemy's fortified position. The first missile struck north of the target—years later, participants in the day's events disagree as to why. The second Hellfire strike was a direct hit against the bunker, and the US forces no longer received enemy fire from that position. Though the Predator was now "Winchester"—it had expended all its ordnance—it was not useless. Brown delegated some of his responsibility as controller to Jones, the Predator's pilot.[62] Jones describes checking French Mirage 2000 fighters into the area and identifying enemy fighting positions. He was even able to use his own aircraft's laser to provide terminal guidance to the French pilots' laser-guided bombs—the same buddy-lase technique that Predator pilots had used in Kosovo just a few years earlier. Brown, the combat controller, would later say, "I credit that pilot, the technology, and that airframe to saving my life as well as the team's and getting the wounded and [killed in action] off that hilltop that day."[63]

Though many associate the Predator solely with high-value targeting operations, from March 2002 on, close air support became an important and frequent mission for Predator crews. One air force colonel who participated in the Predator's development said,

> That night, it was more than an experiment; it was saving American lives. We were a sideshow up until that point in time. People were talking about us, but not as something that was going to be a long-lasting thing. After that, Predator became what it is today. Nobody ever doubted us again.[64]

In the first six months of the US war in Afghanistan, the Predator had conducted persistent intelligence, surveillance, and reconnaissance, high-value targeting, and close air support. In the years that followed, this airplane would be asked to conduct an increasingly wide array of missions previously reserved for traditional pilots and traditional aircraft.

The set of missions widened again in 2010. On January 12, an earthquake that reached 7.0 on the Richter scale devastated Port au Prince, Haiti.[65] Nations from around the world sent aid, and the US Army sent components of its 82nd Airborne Division to manage security and food distribution. US Air Force Predator crews provided real-time full-motion video coverage of food distribution sites and of the roads that traversed the island from east to west.[66] These roads were the backup corridor for food and supplies in case the single runway at the Port au Prince airport became inoperative. But the Predator's role in Operation Unified Response was significant for reasons beyond the specific operation. Until then, the US Air Force had no agreement with the Federal Aviation Administration (FAA) to fly these aircraft in the national airspace. In support of Haiti relief efforts, the Predator crews had to launch from Puerto Rico, which is, of course, US national airspace.[67] The air force required and obtained an "emergency certificate of authorization" from the FAA. On its own, the Predator could not meet the FAA requirement that pilots must be able to "see and avoid" other traffic. The emergency certification required that a qualified pilot would stand on land in Puerto Rico with a pair of binoculars, looking for visual traffic.[68] This person would "see" on behalf of the pilot in the cockpit. This arrangement, crucial to the Predator crews' support in Haiti, foreshadowed the 2012 congressional mandate that the FAA "integrate unmanned aircraft into the national airspace."[69]

Not long after the humanitarian response in Haiti, the Predator community was faced with a new operational challenge. The 2011 North Atlantic Treaty Organization (NATO) air operation in Libya was a different kind of air war from those in Iraq and Afghanistan. In Libya, strike aircraft regularly conducted strike coordination and reconnaissance missions.[70] Unlike the close air support missions of the previous decade, in which pilots endeavored to meet ground force commander intent in close coordination with a controller on the ground, the tactics required in Libya "placed more responsibility for positive identification on aircrew members, and thus required a great deal of diligence and fire discipline."[71] Pilots—of both remotely piloted and traditional aircraft—would now have to identify enemy forces and equipment on their own authority. In addition to strike coordination and reconnaissance as a new mission set, according to some reporting, the operation in Libya required Predator crews to participate in the destruction of enemy air defenses for the first time ever.[72] This slow and relatively unmaneuverable aircraft would now operate inside the threat envelope of the enemy's surface-to-air systems to target those systems.

More recently, the Predator and Reaper have played an important role in the fight against ISIS in Iraq and Syria. In places like Mosul in Iraq and Raqqa in Syria, this meant supporting forces on the ground in complex urban environments. Crews would now be asked to orchestrate strikes near friendly forces and, at times, in crowded cities. In this environment, crews delivered munitions sometimes within twenty-five meters of friendly forces. They became proficient in "urban close air support," and developed multiship operations and tactics. As one pilot explained it to me, "This was a constantly moving scenario. . . . [Friendly forces were] trying to retake a city, so it was street by street, block by block."[73] According to one defense reporter, in the fight against ISIS, the Predator and Reaper became a "cornerstone of the US military's air campaign." Lieutenant General Harrigian, commander of the US Central Command Air Component said in 2017 that the Predator and Reaper "community . . . has been instrumental to the defeat of [ISIS]."[74]

The story of the Predator is not simply the 2001 emergence of a new means of targeted killing. Its history recounts a maturing weapons system in parallel with a maturing community of practitioners who employed it. When I first saw the Predator that summer in 2003, its pilots and sensor operators were still reeling from the addition of the Hellfire missile and the growth from two active combat Predator missions at any given time to four. Over the next decade, it would grow from four active combat sorties to sixty-five active combat Predator and Reaper missions at any given time.[75] If we want to understand the Predator as a model of remote warfare, we must distinguish between weapons systems and missions. The Predator has played an important role in high-value targeting operations, but it has played equally important roles in other missions, too. If remote warfare is going to be understood as a category, it ought to be evaluated across this wide range of relevant missions.

My sweeping history of the MQ-1 Predator sets the table for the ethics discussion that follows. This "odd bird" that inaugurated twenty-first-century remote warfare has been widely misunderstood but has forced us to conceive of an entirely new category of weapons, namely, remote weapons systems.[76] I have tried to distinguish clearly the view of the Predator as a weapons system from the concept of high-value targeting (or targeted killing) as a mission. The Predator was not originally designed to hunt terrorists. It was developed to find and identify heavy mobile weapons and then to direct fighter and bomber aircraft to those targets quicker than previous aircraft had been able to do—to close the sensor-to-shooter loop. Because the enemies the United States would face in the twenty-first century were different from those in the 1990s, the aircraft was adapted to the targeted-killing mission in 2000 and 2001, out of necessity. The distinction between the Predator and high-value targeting is also evident in the wide array of missions to which the Predator and its replacement, the Reaper, have been assigned.

Before going any further, I should mention something about the language that has grown up around these weapons systems in the twenty-first century. For many, the spectral imagery that accompanies the remotely piloted US aircraft is confirmation of their depravity. The Predator was replaced by the grim-sounding Reaper, and their first and best-known munition is called the Hellfire. The Reaper is said to be the first purpose-built "hunter-killer drone."[77] We are likewise told that the software designed to predict the potential damage caused by a munition is called Bugsplat and that this software tool helps estimate collateral damage.[78] Some of these terms—Predator, Reaper, and Hellfire—seem to celebrate death and to put the macabre at the center of remotely piloted aircraft imagery. Others, we are told—for instance, Bugsplat and *collateral damage*—are euphemisms that encourage crews to treat enemy fighters, and even civilians, as subhuman. But how much should we allow this lexicon to color our view of remotely piloted aircraft and their crews?

The language is unfortunate. An external observer will perhaps understandably see this lexicon, dripping with death and darkness, and draw conclusions about the crews and how they approach their violent work. But to think the Predator and Reaper community is responsible for these terms is to ascribe to that group far more power within the air force than the community actually has. Each of these terms has its own genealogy, and the term's association with remote warfare was usually by mere happenstance. For instance, it was the company that built the Predator, not the air force, that insisted on the name Predator. In fact, the company had named two previous aircraft—both of which ultimately failed—by that name before this one became a household name.[79] The *hunter-killer* term is a reference to fighter aircraft in Vietnam. By the end of that war, dissimilar fighter aircraft were paired into hunter-killer teams to target enemy surface-to-air missile systems.[80] And the Hellfire missile was neither designed nor named for the Predator and Reaper. It was an AH-64 Apache helicopter munition long before the Predator inherited it. Indeed, it was originally called the HELiborne Laser FIRE and Forget missile or, simply, the Hellfire.[81]

These terms, taken on the whole, do cast a dark cloud over Predator and Reaper operations. The result of this potpourri of historical terms is an unfortunately grim lexicon that looms over the remotely piloted US aircraft fleet. And if one wants to lay blame, then surely responsibility for these terms falls to the military-industrial complex. But it would be a mistake to allow this lexicon to influence how we think about the crew members. They did not ask for an airplane called Predator, a missile called Hellfire, or mission planning software called Bugsplat. If we want to understand the morality of remote warfare, we would do well to look past these terms to the central moral questions raised by remote warfare. I begin to do just that in the next chapter.

# 2

# RISKLESS WARFARE?

M Y GRANDFATHER WOULDN'T CALL HIMSELF A WARRIOR. HE probably wouldn't have wanted me to call him a warrior either. Like many veterans of the Second World War, he spoke little of his combat experiences. But whatever the threshold for being a real warrior is, whatever risk someone must accept, or whatever actions someone must take to meet the warrior standard, surely my grandfather did enough.

On June 26, 1944, thirty-six bombers of the 455th Bomb Group taxied to the runway on their base in southern Italy. The heavy B-24 Liberators left the ground and lumbered to join the formation overhead. Their destination was the Moosbierbaum Oil Refinery outside Vienna. My grandfather, Lieutenant Frank Randall, was squeezed into the glass nose of an aircraft nicknamed "Snuffy Smith" and assumed his position as the aircraft's bombardier. As the formation began the final leg of the route toward the target, it would be the bombardiers who would control the aircraft. The pilot and copilot physically took their hands off the controls as the bombardier looked through the Norden bombsight at the Austrian ground below. Through a series of knobs and dials, the bombardier would manipulate the airplane's control surfaces to fly it over the target.

The fighting that day was fierce. As the heavy bombers began the bomb run, they were engaged by about 120 German fighters. Single-engine and twin-engine fighters shot through the formation, engaging the bombers with machine guns while evading the returning fire from the B-24s' many gunners. At the same time, large-caliber antiaircraft artillery rounds were trained on the formation's altitude, and flak burst all around the bombers. One of the B-24s crashed headlong into an enemy fighter and burst into flames. Even so, its crew members managed to maintain their position in the formation long enough to release their bombs over the target before plummeting to the earth. Two other bombers were hit by gunfire. They, too, managed to release their bombs over the target before exploding in midair.[1]

Losses were heavy on both sides. Of the thirty-six aircraft the bomb group launched that day only twenty-six would return. Each B-24 had a crew of ten. Ten lost bombers meant that one hundred airmen were either killed or parachuted into enemy territory. The 455th Bomb Group lost more aircraft on that day than on any other day of the war. But the B-24 gunners did their work too, downing thirty-four German fighters. My grandfather survived unscathed and was among those awarded the Distinguished Flying Cross for his actions that day.

Whatever it takes to be counted as a "real" warrior, these men had it. Can a Reaper crew member, seven thousand miles from the death and destruction of war, be considered a warrior?

Seventy years after Moosbierbaum and six thousand miles to the west, Predator and Reaper crews would experience war with such little risk to themselves that it would have been unimaginable to members of the 455th Bomb Group. Instead of rumbling down a taxiway at an airfield previously captured from the enemy, Reaper crews commute into the Nevada desert at odd hours of the night and day. Instead of fighting off enemy aircraft while inbound to the target, Reaper crews command their craft to loiter overhead in uncontested airspace. And instead of maintaining altitude and airspeed in the face of effective antiaircraft fire, the Reaper crews remain safely on the ground on the other side of the globe. Unlike what my grandfather and his 455th Bomb Group faced, the combat risk is not distributed between friendly and enemy aircraft crews. Instead, it is distributed among friendly forces on the ground, enemy combatants, and, tragically, as in all wars, civilians.

If Predator and Reaper crews conduct military operations without facing physical risk themselves, can what they do be really considered war? The difference in physical risk between my grandfather's experience of war and my own is vast. Indeed, the whole history of warfare seems intertwined with physical risk to combatants. Remote warfare represents a drastic departure from this norm. What, we are left to wonder, is the relationship between risk and war? Can riskless war really be war at all? Or, more central to the ethical questions, can a combatant who does not face physical risk be a true warrior?

Though this issue of risk will continue to arise throughout this book, I am interested here in the relationship between risk and the warrior ethos. Some have argued that the warrior ethos depends on physical risk and is therefore unavailable to remote warfighters. These challenges all coalesce around a single question: Can a person be a warrior without facing physical combat risk?[2]

Any discussion of the warrior ethos is fraught in part because its definition is so hard to pin down.[3] Military ethics scholars use the term *warrior ethos* to refer to the internal force constraining the warrior's violence—preventing the warrior's actions from devolving into blind aggression or indiscriminate massacre.[4] But at the same time, most authors seem to suggest that this warrior ethos is forged in the heat of battle, under intense pressure, and, perhaps most importantly, at great risk to the warrior. Remote warfare thus faces a challenge:

Can remote warfare crews cultivate the warrior ethos, and if not, should that be morally troubling?

I show in this chapter that the warrior ethos does have moral value—but that this value is not grounded in the risks a combatant faces in combat. In the many wars throughout history, soldiers have been asked both to take life and to risk death. In many martial contexts, the two are so tightly correlated with one another that we need not trouble ourselves with the distinction. But remote warfare has separated these two elements of the warfighter experience. The remote warfighter takes life but does not risk death. How are we to understand the warrior ethos once these two elements have been separated? If anything, a survey of the academic literature reveals that no two scholars interpret the warrior ethos in exactly the same way. Whatever the warrior ethos is, it is wrapped up in notions of heroism, courage, sacrifice, emotional connection, existential transformation, risk-taking, and what it means to go to war.[5] Most discussions about the warrior ethos and remote warfare fail to account for this complexity. In this chapter, I show how this ethos might apply to combatants who face the possibility of taking life without risking their own.

## THE WARRIOR ETHOS

The warrior ethos is grounded in mythology because, throughout history, war has been romanticized. Young people dream of battlefield heroics and the triumphal return, and when war is remembered, acts of heroism, sacrifice, and love come to the fore. War is seen as an opportunity for honor and glory. This romantic vision of war in the abstract fails to capture the practical realities of war. Some might readily admit that we romanticize the industrialized wars of the twentieth century, but what of "traditional" war? Surely, one might think, when wars consisted in mounted knights on white chargers, glinting swords, and bright shields, war was good, just, and honorable. Surely war was romance. But if the old wars seem more romantic, it is only because we remember the glory and forget the gore.

Oliver Wendell Holmes Jr., a Civil War veteran of the Union Army, was wounded in combat on three separate occasions. When he was just twenty years old, Lieutenant Holmes's unit, the 20th Massachusetts Volunteer Infantry Regiment, exchanged fire with a scout company at Ball's Bluff, Virginia. When his commander directed a charge, Holmes recounted, "I waved my sword and asked if none would follow me." Just then a round went right through his chest— he later found it lodged in his clothes. He fell to the ground and lost consciousness. During his first night in the crude field hospital, his prognosis was poor. He later wrote to his mother that he had "made up [his] mind to die" on that night—though, of course, he lived. Holmes was ultimately transported back to his parents' home in Boston to recover. One of Holmes's closest friends during the war, Henry "Little" Abbott, wrote to his own father after Ball's Bluff, calling Holmes "a devilish fine fellow and a devilish brave officer."[6]

Looking back on his military service, Holmes would distinguish between mythologized war as we imagine it and war as it is. In 1923 he wrote to a friend about this distinction:

> It would be easy, after a comfortable breakfast, to come down the steps of one's house, pulling on one's gloves and smoking a cigar, to get onto a horse and charge a battery up Beacon Street, while the ladies wave handkerchiefs from a balcony. But the reality was to pass a night on the ground in the rain with your bowels out of order and then, after no particular breakfast, to wade a stream and attack the enemy. That is life.[7]

In what is perhaps his most famous speech, given three decades after the war, the Supreme Court justice described the realities of warfare: "War, when you are at it, is horrible and dull. It is only when time has passed that you see that its message was divine."[8] Holmes spoke from his personal experience of what scholars have often told us: we mythologize the wars of the past.[9] Though war had been made industrial in Holmes's time and digital since the 1990s, death and destruction remain enduring facets of war.

The warrior ethos is mythological in another sense, too. The Western understanding of who the warrior is and what the warrior ethos means begins with Homer and his Achilles. Achilles, son of a goddess, is famously unassailable but for his heel. If, as some have argued, Achilles is the archetypal warrior and his warrior spirit the original template for the warrior ethos, what hope can mere mortals have of being admitted into his class?

The traditional warrior we see in much of the writing about remote warfare is as mythical as Achilles. We are sometimes presented with an image of war in which the warrior is at the center—war exists as a transformational experience for the warrior.[10] This self-referential crucible of the mythology has enticed young people into combat throughout human history. A young Alexander Hamilton, for example, raised on a steady diet of tales of honor and glory in Europe's colonization of the West Indies, daydreamed about the glory war promised. He once lamented his lot in life to a childhood friend and spoke of war as his only hope of overcoming it: "[I am] confident, Ned, that my Youth excludes me from any hopes of immediate Preferment nor do I desire it, but I mean to prepare the way for futurity. . . . I wish there was a War."[11] Only a war could secure for him the prestige and position he sought in life. Hamilton wanted not merely a chance to test his mettle but also the chance to show others that his mettle could withstand the test. The outward honor and glory that he could win in war—and only in war—would be the means by which this destitute orphan would secure a place in society, in government, and, ultimately, in history.

This desire for recognition for having overcome war's challenges is much older than Hamilton. As one historian writes, the martial attitude in ancient Rome was perhaps even more pervasive among young men. "There was intense competition among [the young men] for glory: each one of them hastened to strike down an enemy, to climb the rampart, and to be seen doing such a deed."[12] Here too, what is at stake for the young men is not the mere act or even the acceptance of personal risk to self. Instead, the aim is to complete the act at great

risk under the watchful eye of an adoring public—to be seen doing such a deed. The honor and glory described in these passages are
not just internal character virtues but the external praise one seeks to
win. The warrior does not merely survive the self-referential crucible
but is seen surviving it.

In the ancient Greek city-state, one's identity as a man was inseparable from one's identity as a soldier. For men of every age in
the Greek city-state, manhood was defined in terms of his ability
to stand his post in the phalanx against the onslaught of opposing
spears. Military historian Victor Davis Hanson writes that the ancient Greeks "understood that the simplicity, clarity, and brevity of
hoplite battle defined the entire relationship with a man's family and
his community, the one day of uncertain date that might end his life
but surely gave significance to his entire existence."[13] A man's identity
as a member of the community was inseparable from his identity as a
member of the phalanx.

Similar sentiments run right through to the wars of the twentieth
century. J. Glenn Gray, philosopher and veteran of the Second World
War, writes, "I confess that in my adolescence and early manhood,
before World War II, I longed for one more war, in which I might
participate. Though I never spoke of such a wish, and regard it today
with considerable dismay, I cannot deny that it was an important
part of the aspirations of my youth. And I have no reason to believe
that my case is unique or singular."[14]

US warfighters describe this longing for their first encounter with
combat as "seeing the elephant." The phrase is at least as old as the
nineteenth century, though some claim it is much older. US soldiers in the Mexican-American War spoke of seeing the elephant to
"describe the letdown they experienced after their expectations of
glory."[15] With recent combat experience in the Mexican-American
War—to say nothing of the violence he would later witness in the
US Civil War—a young lieutenant Ulysses S. Grant made the same
point to his wife, Julia: "Wherever there are battles a great many
must suffer, and for the sake of the little glory gained I do not care to

see it."[16] One cannot understand what it means to go to war until one has been there and seen it up close. And this adventure, like all adventures, calls to young people in every generation. But the promise of romance and glory is rarely fulfilled.

The trouble with defining war or the warrior ethos in terms of the honor and glory of the mythology is that war rarely behaves according to mythical expectations. The First World War provides the clearest example of this phenomenon. It followed Industrial Revolution developments that mechanized warfare, and it took place during a time characterized by fading social mythologies throughout Europe. Though many expected the war, for example, to be an honorable and glorious endeavor, the lived experience of the frontline soldiers tells a different story. One French soldier in the trenches of the First World War echoes Grant's epiphany from some seventy years earlier: "If these notes should reach any one, may they give rise in an honest heart to horror of the foul crime of those responsible for this war. There will never be enough glory to cover all the blood and all the mud."[17]

A generation later, Europe would again experience the tragedy of war. Though there are innumerable stories of moral virtue and of sacrificial love on far-flung battlefields, there is no question that the war was a tragedy. "A walk across any battlefield shortly after the guns have fallen silent," Gray writes, "is convincing enough. A sensitive person is sure to be oppressed by a spirit of evil there."[18] Even a just war is tragic.

This brief survey is neither exhaustive nor indicative of every war veteran's experience. It is meant only to show that the tension between expectations of glory and honor on one hand and the cold, devastating realities of war on the other has been recognized by combatants throughout the long history of political conflict. We might be tempted, nevertheless, to ground the warrior ethos in battlefield honor and glory. And in so doing, we might quickly find that remote combatants—Predator and Reaper crews—would be excluded from that ethos. But the problem with basing the warrior ethos on the mythology of war's honor and glory is that such a fragile ethos will be

suffocated by "all the blood and all the mud" of war in actual practice. This is perhaps why, for so many fighters, the mythology-based
romanticism of war cannot survive artillery barrages and dead comrades in arms. The glory of warfare is visible only through the lens
of mythology. This view is, at best, an appropriate means of looking
back at the battles the warfighters have survived. But it will be of little help to those who must look to battles that lie ahead.

In his book *The Warrior Ethos*, philosopher and military ethicist
Christopher Coker devotes considerable time and space to Achilles
and calls him "that archetypal hero" and "the sine qua non of the
Western Warrior."[19] In one sense, it is easy to see why Achilles is often viewed as the warrior archetype. Even our contemporary descriptions of what warriors must do and what kind of attitude they must
adopt echo this ancient warrior. Surely Achilles would understand
the "fangs-out" fighter pilot culture in the US Air Force and US Navy.[20] Achilles had the kind of bias for action that has long been associated with the US Marines, and he was the consummate expert at
enhancing close-combat lethality in complex terrain long before the
2018 US National Defense Strategy articulated that goal for its own
forces.[21] But as the military ethicists who study Homer are quick to
point out, Achilles is no paragon of virtue.

Achilles is a complex character, and even if my space to discuss
him here were unlimited, I lack the skills to do him justice. But we
need to spend very little time with Achilles to recognize behaviors
and character traits that we might not want our own combatants
to emulate. Though, as Coker puts it, Achilles "continues to define
the parameters of the heroic," his heroism lacks the sense of other-
centeredness that is required of modern professional militaries.[22]
Achilles's heroism and his great battlefield deeds are motivated not by
a desire to obtain victory on behalf of a political community but by
his own pursuit of honor and glory:

> While Achilles is undoubtedly heroic, he is also cruel. He sees war in no
> other light than the scope it provides for his own heroism. His greatness

lies almost wholly in his courage and force of will. He has little humanity and even less imagination. Today, our heroes have to be fired by more
than personal ambition.[23]

Homer's *Iliad* opens not with Achilles's wisdom, mercy, charity, or
love of country but with his wrath. Achilles's rage against a slight to
his personal honor is what drives much of the plot in the first place.[24]
The Greek expeditionary force lands at Troy without their best
warrior because Achilles's personal honor has prohibited him from
supporting Agamemnon. Though Achilles might be the archetypal
warrior, he is not a suitable role model for modern warriors.

By Plato's time—still some twenty-five hundred years before our
own—the flaws in Achilles's ethos had become apparent. As Plato's
Socrates mulls censorship in the ideal city, he wants to edit the self-
ishness and greed out of Achilles's character. In Book 3 of *The Republic*, Socrates says, "Nor will we suffer our youth to believe that
Achilles . . . [was] affected with two contradictory maladies," greed
and arrogance.[25] Plato's censorship of Homer is intended to reinvent
Achilles not as warrior motivated by self-aggrandizement but instead
as servant of the political community, engaging in war on behalf of
others.[26] Achilles is out for himself. But surely what we're after in a
warrior ethos is something less self-centered. In our own world, contemporary Western states want their militaries not to be filled with
privateers but to be made up of public servants.

One way to describe Achilles's limitations as a role model for the
warrior ethos is by comparing him with Hector. Achilles is not just
a warrior. He is all warrior. We get no sense of his humanity outside
of war—if he has any at all.[27] If the warrior ethos means fighting
not for oneself but for the good of the community, then surely it is
Hector, rather than Achilles, who ought to serve as the archetypal
warrior. Perhaps Hector can serve as a role model because he is both
a good warrior and a good person, but if so, we end up back at the
doorstep to the original question: What is the moral value of the
warrior ethos?

Surely Achilles and Hector, as warriors, had cultivated an ethos that enabled them to face great dangers and submit themselves to great risk—ultimately, mortal risk in both cases. But exposure to risk is only part of what Achilles was willing to do for his own honor and glory and only part of what Hector was willing to do for his fellow Trojans. Achilles and Hector not only face risk but also take lives. The warrior ethos as embodied in Achilles and Hector is about killing as much as it is about dying. And on this point, Achilles and Hector stand in stark contrast with one another. Both are brave; that was never in doubt. But how does each of these ancient warriors approach the act of taking life? Achilles kills out of his own interest in honor and glory. Hector kills to defend his home, his family, and his fellow citizens. But a warrior ethos that focuses only on risk will neglect this important difference. This relationship between the warrior and killing is easily overlooked in part because throughout military history, acts of killing are often accompanied by the risk of death. In remote warfare, the physical risks are one-sided, and this forces us to distinguish between these two facets of the warrior ethos: risk of death on the one hand and acts of violence on the other.

In an important sense, the use of force is even more central to the warrior ethos than the risk of death is. Theoretically, we can see the primacy of the use of force if we compare the various groups who are affected by war. The traditional warrior faces risk and commits violent acts. The remote warfighter does not face risk but does commit violent acts. But what of the noncombatants who are affected by war? What about refugees who are forced from their homes and other noncombatants who lose loved ones or whose property is destroyed in the fighting? In some cases, civilians suffer far more than combatants do—or at least the suffering is much more widespread among noncombatants. If the warrior ethos is defined primarily as risk to self, then surely remote warfighters are excluded from the warrior ethos. But if combat risk is the only factor relevant to the warrior ethos, then noncombatants are warriors, too. And this can't be right. There must be more to the warrior ethos than the risk of death in war.

A second reason to define the warrior ethos primarily by its relationship to causing harm and not just facing risk is that an outsized focus on facing risk will result in a bizarre glorification of some warriors over others. If the warrior ethos is defined primarily in terms of one's acceptance of physical risk, then presumably, accepting greater risk makes one more of a warrior. Or at the very least, those who are more willing to risk their lives are more qualified for the class of warrior than are those less willing. Perhaps, in this view, the fighter pilot is a warrior but not as much a warrior as the artillery soldier is. The armor soldier in a tank is not as much a warrior as the infantry marine who stands with two feet on the ground. In this light, perhaps the remote combatant is not a warrior at all.

Interservice rivalries aside, this hierarchy might seem plausible on the surface, but it fails at the extremes. If the warrior ethos is defined solely in terms of risk, and if greater risk means greater warrior status, then at the pinnacle of warrior elites is not the infantry soldier but the suicide bomber.[28] Many observers will no doubt reject this conclusion. But if so, we must come up with a better account of the relationship between the warrior ethos and risk. The caricature of the warrior ethos as defined solely by the risk the warrior accepts, then, must be rejected for a more nuanced view.

At the same time, risking one's life—or at least the willingness to risk one's life—does seem foundational to what it means to be a warrior.[29] Maybe it is not risk itself, but the *willingness* to face risk on behalf of one's political community that is a fundamental facet of the warrior ethos. But are remote warfare crews unwilling to face risk? They have undergone education and training that is similar to their peers. They have put themselves under the legal authority of their commanding officers and, ultimately, the civilians who control the military. Pilots and sensor operators have raised their right hands and sworn an oath to support and defend the Constitution—the very same oath their traditional warfighter counterparts have sworn.

My wife and I recently visited the beaches and the American Cemetery in Normandy. It was as powerful an experience as everyone told us it would be. We stood on the spot at which the liberation of Europe began, and we walked among the more than nine thousand tombstones commemorating the Americans who were killed liberating it. In a line to board a bus, I found myself standing near an older gentleman in a WWII Veteran cap. We got to talk briefly. On June 5, 1944, Bill had started the night shift at a Baltimore newspaper. In the very early hours after midnight, on June 6, all the teletype machines came alive with news of the invasion. It had been his job to wake the editors in case of breaking news—and this was breaking news indeed. He told me that because of his job and the night shift, he must have been one of the first people in America to learn of the D-Day invasion. That very day, at seventeen years old, Bill got his mother to sign his enlistment form, and he joined the navy. By the time his training was complete and his ship was headed for Europe, the war was almost over. He joked, though, that he got to the Pacific theater just in time to earn his WWII Veteran hat.

We can say that Bill faced no great risk during the war. But there is no question that he was willing to face risk. He volunteered and swore the same oath sworn by the more than nine thousand men and women whose bodies are now interred in Normandy.

I have met Predator and Reaper pilots who made the transition from traditionally piloted aircraft for several reasons. Some got airsick. Some wanted to have more of an operational impact and to be home with their families more than they could while flying heavy aircraft around the world three hundred days a year. I have met far more who sought careers in traditional aircraft but who, for any number of reasons, were assigned to remotely piloted aircraft by the Air Force Personnel Center. But I have never met a remote crew member who volunteered to join the military, swore an oath to support and defend the Constitution, and then opted for Predator or Reaper to avoid the risks of combat. If the minimum requirement for attaining the warrior ethos is the willingness to accept risk, then by taking their oath of office, military members in general have expressed that willingness—regardless of the career field in which they ultimately find themselves.

This willingness to face risk does not, by itself, account for the warrior ethos. Besides one's willingness to risk one's life, the ethos also depends on one's willingness to take the lives of others. The warrior ethos must relate to killing as well as dying. This theme arises anecdotally and theoretically. In his *Band of Brothers*, Stephen Ambrose describes the success of Easy Company's training: "They were prepared to die for each other; more important, they were prepared to kill for each other."[30] Karl Marlantes, a veteran of the US war in Vietnam, likewise casts the role of the warrior not just in terms of risk but in terms of what the person will do to enemy combatants: "You can't be a warrior and not be deeply involved with suffering and responsibility. You're *causing* a lot of it. You ought to know why you're doing it. Warriors must touch their souls because their job involves killing people. Warriors deal with eternity."[31]

Military historian Victor Davis Hanson describes the relationship between the ancient Greek warrior and acts of killing this way: "A physical and psychological exertion of energy was required by men who killed with hand tools and then watched one another struggle, bleed, and go down beneath their feet."[32] The phenomenon is as old as war. Achilles and Hector face death, but they also take life.

What are we to make of this warrior ethos that is grounded in the willingness to risk one's life as well as the willingness to take the life of another? Various authors have insisted that the ethos should be based on mutual and reciprocal risk. The warrior stands and faces an opponent. Each holds the other at risk. If Achilles and Hector meet on the battlefield, both risking death and both threatening the life of the other, perhaps the warrior ethos they embody is better described by the duel.

## WAR AND THE DUEL

One of the common arguments against remote warfare insists that the armed conflict of a war must be like a duel.[33] This argument seems obvious and might be taken for granted. But it is a dangerous proposition.

The argument against remote warfare is simple and significant. War is a duel. Remote warfare operations are not like duels, because one side faces physical risk and the other does not. Therefore, so the argument goes, remote warfare operations cannot be war; nor can remote warfare crews be warriors. The argument is based on the work of a giant in the history of military thought, the nineteenth-century war theorist Carl von Clausewitz. In an attempt to define war, Clausewitz includes among his opening lines "I shall . . . go straight to the heart of the matter, to the duel. War is nothing but a duel."[34] Since Clausewitz remains one of the most authoritative voices on the nature of war, this argument looks like an open-and-shut case: war is a duel. But even though the argument has strong intuitive appeal, the conclusion isn't so simple.

When combatants meet in battle and try to kill one another, they do behave, at least in some respects, like opponents in a duel. Each holds the other at risk. Each wishes not to be killed, yet each has full knowledge of the risks of being killed. The relationship between the field of honor on which duels are fought and the field of battle on which wars are fought is also closely related to the martial virtues that I will examine further in Chapter 6.[35] Many of the martial virtues—courage, honor, and mercy at least—are as relevant to the duel as they are to battle. The European and American duels of the eighteenth and nineteenth centuries are indeed reminiscent of idealized warriors who stand and face the enemy throughout the history of war. "The best and finest thing," the Spartan poet Tyrtaeus wrote in the seventh century BC, is when a young man, "feet planted firmly, stands relentlessly in the front ranks and pays no thought at all to shameless flight."[36] The duel looks like a ready-made analogy for traditional battle in war. The young man stands, feet firmly planted, squaring off against his enemy. The remote warfare pilot, though, does not stand and face the enemy. The pilot is not exposed to risk or to threat. There is no need for the physical courage required to face such a threat. And at thousands of miles removed from the target, the remote pilot does not engage the enemy in a contest.

This is, perhaps, the most disconcerting thing about remote warfare. Rather than meeting one's enemy face-to-face, the Predator or Reaper crew follows the adversary from a distance. To avoid civilian casualties, crews will sometimes follow a high-value target for extended periods of time—always keeping watch, always tracking the target. The French philosopher Grégoire Chamayou, in his *Theory of the Drone*, describes remote warfare operations this way:

> Hunting is essentially defined by pursuit. . . . Combat bursts out wherever opposing forces clash. Hunting, on the other hand, takes place wherever the prey goes. As a hunter state sees it, armed violence is no longer defined within the boundaries of a demarcated zone but simply by the presence of an enemy-prey who, so to speak, carries with it its own little mobile zone of hostility.[37]

When the United States conducts remotely piloted aircraft attacks outside areas of declared hostilities, according to Chamayou, it is allowing the means and methods of warfare to change from combat to the hunt. When adversaries such as Baitullah Mehsud, Anwar al-Aulaqi, or Abu al-Khayr al-Masri are tracked, targeted, and killed, is it warfare? Or is it something different? Chamayou is not the only one to recognize this relationship between remote warfare and hunting. Dave Blair and Karen House have written in defense of remote warfare crews. They describe a Reaper sensor operator's ability to anticipate where a target will go in the final moments of the missile's flight. They call this ability a "hunter's empathy."[38] And if this new kind of warfare is anything like a hunt, then it has very little in common with the duel. And if it is not like a duel, can it really be war?

Despite the superficial appeal, these inclinations toward war as a duel are misplaced. Although the fields of honor and the fields of battle do share some common attributes, it would be unwise to base the concept of battle on the duel or to ground a warrior ethos in the dueling fields. The first and most obvious reason to reject these arguments is that they misconstrue Clausewitz's claim. Clausewitz did not say that each individual *battle* is like a duel. Nor did he say that each combatant engages the enemy the same way a duelist engages an opponent. Clausewitz was not writing about individual battles but was describing war on the whole: "War is nothing but a duel *on a larger scale*. Countless duels make up war, but we can form a picture of war as a whole by imagining a pair of wrestlers. Each wrestler tries through physical force to compel the other to do the wrestler's will; the immediate aim is to throw the opponent to make the rival incapable of further resistance."[39]

Though Clausewitz's contributions are many, perhaps his singular best-known observation is his statement that "war is merely the continuation of policy by other means."[40] This conception of war as policy has embedded itself even into US military doctrine.[41] The US government's instruments of power consist of diplomatic, intelligence, military, and economic means. The military is one means by which the US government can achieve its policy ends. War is relevantly like a duel only when taken as a whole. When conceived at this strategic level, each side seeks to bring its opponent—that is, the political community or communities with which it is engaged—to the point that it or they must capitulate. War is policy by means of violence. War is a duel. But the individual battle is neither policy nor duel.

Moreover, Clausewitz openly rejected at least some facets of the duel as anathema to the battles that constitute war. The European practice of dueling was seen as a continuation of the combat practices of medieval knights.[42] According to the code of Thomas Malory's knights in *Le Morte d'Arthur*, the chivalric tradition on which the duel was based, the hallmarks of shameful and therefore less-than-knightly violence were surprise and subterfuge.[43] The duel was designed to avoid surprise and subterfuge. Imagine a duel in which one participant attempted to defeat an opponent through surprise—lying in wait on the path to the appointed spot and shooting at the foe through the trees. The ambusher might indeed kill his opponent, but he has not participated in a duel. The duel, like the chivalric code of the medieval period that inspired it, rejected any attempts at surprise or deception—hallmarks of admirable military thought in the centuries of war that have followed.[44]

Modern history—as well as ancient history—is rife with examples of subterfuge and surprise in war, and none of the examples seem to challenge the long-standing conception of war or of the warrior ethos. George S. Patton's inflatable tanks and fictitious First US Army Group were designed intentionally to mislead German defenses about the ensuing D-Day invasion. Robin Olds's "Operation Bolo" in Vietnam disguised air superiority F-4 Phantoms as tactical bombers—a successful bid to lure enemy fighters out against what they believed to be less maneuverable and less defensible F-105 Thunderchiefs (Thuds). Norman Schwarzkopf's "Left Hook" feint in Operation Desert Storm and, for that matter, the Greeks' wooden horse at Troy must also be rejected as something other than real war.[45] Each of these examples was a successful attempt to deceive the enemy. Supposedly unfair advantages of this kind have no place in the medieval chivalric code or in the duel that followed it. If Clausewitz really thinks that individual battles must be like duels, then surely he would reject these attempts at surprise and deceit as something other than war. But he rejects none of these efforts. In fact, he extols the value of deception in war.

Long before *On War* was posthumously published (1832), Clausewitz summarized his theory of the principles of war in a letter to his pupil Crown Prince Friedrich Wilhelm in 1812. Many of his principles from that letter abide in military doctrine to this day. In this earlier treatise, Clausewitz advises his pupil that "surprise [is] ranked as a principle in its own right. . . . Surprise was best achieved through secrecy and speed of execution. . . . Surprise owed itself to at least some degree of deception or use of a ruse."[46] Surprise, deception, and secrecy are not just acceptable in Clausewitz's understanding of the battles that make up a war but are celebrated as crucial to effective warfighting. Clausewitz's dueling metaphor should be taken only as far as he intended. War is like a duel in that two political communities square off against one another, each attempting to coerce the other to certain behaviors.

There is no reason to think Clausewitz would have wanted us to extend this metaphor intended for war to every battle.

Clausewitz does, in fact, comment on the remote warfare of his day. He seems to have no trouble at all accepting it as part of real war:

> Weapons with which the enemy can be attacked while he is at a distance are more instruments for the understanding [than are weapons that bring the combatants into closest contact]; they allow the feelings, the "instinct for fighting" properly called, to remain almost at rest, and this so much the more according as the range of their effect is greater.[47]

Clausewitz goes on to argue that these "two modes of fighting" are complementary. The purpose of hand-to-hand fighting is to drive the enemy from the field of battle, while the purpose of firearms is to destroy the enemy's armed force. But in spite of the differences in psychological and physiological effect on those who employ these early remote weapons, in his discussion, Clausewitz makes no reference to the rule of reciprocity many have ascribed to him.[48]

If we are to take seriously Clausewitz's dueling analogy, we must distinguish between war as a political community's policy instrument and the individual acts of killing and dying on the battlefield—acts that partly constitute the war. As we will see, even if Clausewitz is right and war is a duel, battle categorically is not.

There are reasons independent of Clausewitz to hesitate in adopting the duel as a basis for the warrior ethos. The duel concept suffers from many of the same limitations we found in Achilles. This Greek warrior "looks to his existential self" and has an "obsession with self—with individual pride and individual reputation."[49] These character flaws in Achilles can be found at the very heart of the duel. For instance, Achilles's character flaw is most recognizable when he prays for a peculiar kind of victory. He did not pray that his side would defeat its opponents and win the war—he was indifferent to winning and losing at the policy level. Instead, Achilles asked Zeus, Athena, and Apollo that "not one of all the Trojans might escape death, nor one of the [Greeks], but that we [two] might avoid destruction, that alone we might undo the sacred coronal of Troy."[50] Achilles prayed that everyone would die—everyone on both sides—except for him and his closest friend, so that they could receive all the glory and honor.[51]

Likewise, duelists in the eighteenth and nineteenth centuries were not in the least concerned with winning and losing, at least not in any modern sense, but with personal honor and glory. According to one history of the duel, "all that counted was the fact that the two opponents braved a possibly fatal encounter, thus demonstrating that they placed greater value upon their 'honour' than upon their lives."[52] In fact, losing the duel had nothing to do with being shot or killed but was related to abandoning one's own personal honor. Being killed in a duel was, if not a kind of victory, then at least a kind of success. One stood one's ground in the face of great risk and affirmed one's own courage and honor even if death had been the eventual result.

This focus on honor rather than victory in the martial sense is affirmed in the Hamilton and Burr case. After the duel in which Aaron Burr killed Alexander Hamilton, frequent accounts appeared in the press, claiming that Burr had spent the months leading up to the duel "engaged in intensive target practice" and that he "had been in the constant habit of practicing with pistols."[53] Whether the rumors are true or not, the social implication is clear. To prepare rigorously for the duel—to train for it, to try in fact to kill one's opponent—was to violate the neo-chivalric code. To engage in target practice was to fight dirty. Years later, Burr befriended Jeremy Bentham, a major figure in the history of moral philosophy and often cited as the founder of utilitarianism. Bentham wrote, "He gave me an account of his duel with Hamilton. He was sure of being able to kill him, so I thought it little better than murder."[54] For Bentham, Burr was little better than a murderer not because he participated in the duel, for Hamilton is equally guilty of this charge, but because he prepared for it and prioritized killing his opponent over securing his honor. Insofar as duels are affairs of honor, some observers would say that Burr lost, despite killing Hamilton, because Burr had behaved dishonorably.

Just as Achilles's rage was incited over a slight against his personal honor, "duels never took place without personal cause. . . . In general, such conflicts were triggered by an insult, an affront to aristocratic and knightly honour, or whatever the party concerned considered such honour to be."[55] The warrior ethos cannot be built on the duelists' ethos, because war is not a private affair. The duel resides at the violent end of a spectrum of interpersonal conflict. War resides at the violent end of a spectrum of political conflict. The duel exists for the duelists; it is the means by which they will have satisfaction against a foe who has publicly wounded their pride. But war does not exist for the warrior; the warrior exists for war. War can be, and often is, a transformative experience for those who participate in it.[56] Many might look back on their military service and recognize it as the most important time of their lives—a formative period in their process of becoming who they are. But although war can produce powerful experiences for the individual, these would be terrible reasons to start a war. Even if honor and virtue can often accompany war, we would be fools to create the conditions of war simply to find honor and virtue. Honor and glory are the motivating force behind the duel, but they fail as a motivating force behind war. For the duelist, defeating an opponent on the field of honor is a secondary concern. For the military member, defeating an opponent on the field of battle is the primary concern. Honor and glory might indeed accompany violence on the battlefield, but honor and glory are insufficient to justify violence. As J. Glenn Gray puts it, "those who enter into battle, as distinguished from those who only hover at its fringes, do not fight as duelists fight. Almost automatically, they fight as a unit, a group."[57] For the ethos in question to be a warrior's ethos, it must take seriously the corporate, political, and collective nature of war. The duelist knows nothing of these things.

## THE NATURE OF WAR

Remote warfare is also seen as such a shift in the nature of war that it violates the ancient standard in which combatants on both sides impose risk and expose themselves to it. For thousands of years, the argument goes, the one constant in the continuum of military technological development is mutual, reciprocal risk between opposing combatants. Remote weapons such as the Predator and Reaper—and perhaps others—disrupt this continuum. The mutual imposition of risk between combatants dissolves, and this new kind of remote violence is something other than war.

Military technological development does take place generally on a continuum over time, and any line of demarcation between "real war" and other non-war violence will be blurry at best. That is, if developments in military technology can shift the nature of war or challenge the warrior ethos, how confident can we be that remote warfare—and not some previous development—is what generates such a shift and poses such a challenge? This line-drawing problem poses a significant challenge for many of the ethical arguments against remote warfare.

We should be cautious about line-drawing claims for a few reasons, not the least of which is that many such claims have been made in the past. Numerous developments in military technology have been heralded as decisive moments that have changed the nature of war.[58] Many, though not all, of these developments took place in the late nineteenth and early twentieth centuries and were nevertheless followed by two gruesome wars that spanned the globe.

The Wright brothers, for example, believed their development and its military application "would make wars practically impossible." But in 1917, Orville Wright described how these hopes had been dashed:

> When my brother and I built and flew the first man-carrying flying machine, . . . we thought . . . that no country would enter into war with another of equal size when it knew that it would have to win by simply

wearing out its enemy. Nevertheless, the world finds itself in the greatest war in history.[59]

In 1904, just a year after the Wright Flyer's first flight, Jules Verne prophesied about military technology in a different domain, when he claimed that because the submarine could target surface ships with impunity, it "may be the cause of bringing battle to a stoppage altogether. . . . War will become impossible."[60] The same has been said of dynamite, the machine gun, and the atomic bomb.[61] These assertions—that technology would be so potent as to put an end to war—are not quite the same ones being made about remote warfare. But the two sets of claims do have common elements. Though they have taken place over a century, each assertion entails that a particular technological development will have a decisive effect on the nature of war. It will make wars too costly, too long, too bloody, or, in the remote warfare case, too asymmetrical or too riskless. These predictions, as we know all too well, were never fulfilled. War continued despite these technological advances. As military technology developed, it modified our understanding of what modern war is. It did have an effect on the character of war, but not the promised, decisive effect. We should, therefore, maintain some skepticism in the face of claims that a new application of technology—in this case, remote warfare—has changed the nature of war.

In addition to the preceding claims about individual technological developments, there is also much to say about entire periods of technological development. The Industrial Revolution had its effect on warfare as early as the US Civil War but was fully manifest in the First World War. Many of the very same criticisms frequently directed at remote warfare were levied against military technology in that conflict as well. Some of the criticisms of weapons of the First World War look as though they might have been lifted from our own contemporary discussions of remote warfare. One young German soldier wrote, "It's the nature of the fighting which is repulsive . . . wanting to fight and not being able to defend yourself! Attacking, which I imagined would be so glorious, what is it but the urge [to rush] forward to the next shelter against this hail of murderous shells? And of the enemy sending them over, not a sign!"[62]

The German soldier's predicament ominously foreshadows the dilemma facing those targeted in remote warfare strikes. Many commentators have stressed that the risk asymmetry between the Predator or Reaper crew and their adversaries is the crucial element that makes remote warfare of the twenty-first century distinct from all its predecessors. In Predator and Reaper missions, the crew is "removed from combat."[63] And those targeted in remote warfare strikes are "unable to respond in the traditional . . . manner" and are "unable to strike back, or at least to strike back justly"; another writer argues that "an insurgent confronted by an army of [remote weapons] no longer has any target to attack."[64] But the German soldier under a hailstorm of artillery fire faced the same constraint a hundred years earlier. This similarity across a century of rapid developments in military technology emphasizes both the stubborn nature of war that refuses to bend under novel pressures and the inherent difficulty with this kind of drawing of lines.

For instance, the generation of fighter and bomber aircraft and crews that preceded the Predator might be equally subject to the charge that they have violated what it means to be a real warrior engaged in a real war. Dos Gringos, the guitar band made up of two air force fighter pilots—and their song "Predator Eulogy"—has drawn the attention of those who see remote warfare as changing the rules of the game.[65] At a time when fighter pilots were being involuntarily moved from their fighter cockpits to the Predator's ground-based cockpits at Creech, the Dos Gringos chorus celebrates the enemy's victory:

> *They shot down a Predator*
> *That's one less slot for me.*[66]

But if we're going to accept Dos Gringos as an insight into air force culture and its rejection of new technology, maybe we should look at the band's broader corpus. In "JDAM Blues," the singers lament the advent of the other recent technological developments—in particular, the Joint Direct Attack Munition GPS-guided bomb, a weapon that has become a staple in US fighter operations. The precision and ease of employing GPS-guided bombs, according to the lyrics, robbed the fighter pilot of some swagger and verve.[67]

> *Not like the old days, back in Vietnam, when they were rollin' in,*
> *droppin' dumb bombs . . . flack so thick you could walk on it.*
> *Now it's just droppin' . . . GPS-guided bombs*
> *from 30-something thousand feet.*

These Viper pilots, with the caricatured bravado for which the duo is well known, are *complaining* that they can now drop more-precise weapons with greater ease from safer altitudes! Dos Gringos does decry the Predator, but the criticism takes its place alongside the criticisms of their own aircraft and its technological advances. And what, we might ask, would previous generations of fighter pilots have said about today's stealth, supersonic, high-performance marvels? Perhaps they might see the F-16's suite sensors and high-tech weapons as unfit for a real warrior.

More recently, a fighter pilot from a different airplane raised these kinds of questions in a different medium. A-10 pilot Colonel Ryan Hill questioned his own martial experience in a poem that includes the following stanzas:

*Generations have voiced through pen and art, the glorious horrors of war*
*Stories are told, some without words, of men changed down to their core*

> *I've been and seen and felt and feared, but my story is not the same*
> *With a different view of the battlefield, I'm wary of what I claim*

> *I have dented the earth and bent the air inside the enemy's door*
> *But I cannot help but ask myself, Have I ever been to war?*

*I've topped the heights and flung my craft*
*into valleys in the black of night*
*But the intimate pain and guilt in death remained outside my sights*

*I've squeezed the trigger that ended men's*
*lives but did not witness the gore*
*So again I have to ask myself, Have I ever been to war?*[68]

The question many have asked of remote warfare is the one Hill asks himself. What amounts to real war? What is the minimum threshold for seeing the elephant? I do not mean to suggest that the Reaper pilot and the fighter pilot face the same levels of physical risk. Instead, I suggest only that if we attempt to define what warfare is in terms of the risk a warfighter faces, then we discover some difficult questions about where the line ought to be drawn.

When I was a lieutenant with only about a year of experience in the Predator, I volunteered to deploy to Afghanistan. I was motivated by the fact that, though I had participated in combat operations from Creech Air Force Base in Nevada, I hadn't really been to war. I wanted to do my part—I wanted to see the elephant. For me, at that time, I thought the important distinction was between physically going to war and staying home. In Afghanistan, I learned of other distinctions, for instance, between those who leave the base (e.g., on combat patrols or convoy operations) and those, like me, who remained "inside the wire." Then there is a finer distinction for those who operated outside the wire: Who was shot at, and who wasn't? The army went as far as to formalize this distinction. In an effort to recognize soldiers from units and career fields that aren't typically associated with combat, it introduced the Combat Action Badge to recognize "soldiers who personally engaged, or are engaged by, the enemy."[69] Though the army seems to have had the best of intentions, it is not uncommon to hear soldiers bemoan officers and noncommissioned officers who jump on patrols not exactly from a sense of duty or loyalty but to meet the minimum criteria for the

Combat Action Badge. In one army captain's account, his patrol had taken sporadic but accurate fire. His Afghan National Army team-mates returned fire with mortars, and "that was pretty much that." While he was filling out the after-action report, he remembers, an officer approached him:

> An objectively worthless turd of an infantry Lieutenant . . . walked into my office and actually asked me if we had taken fire that day. That's right, he had slept through his first (and only) brush with combat but was awarded a badge anyway because regs are regs.[70]

Drawing a line between which experiences ought to count as real war and which ought not to count is an impossible task. There will always be someone worse off. There will always be someone who is given a more difficult mission and who faces more risk.

We tend to overemphasize the role of reciprocity in war. The claim that individual battles must be characterized by reciprocity seems at first to cast doubt on the legitimacy of the remote warfare crew. They fight from a distance and generate an asymmetry. That was also the case with some of the fighting in the two world wars, because of tech-nological developments such as the machine gun, aerial bombard-ment, and artillery fire. If we are to accept the assertion that remote warfare cannot be war, then we must also accept that vast swaths of what many would consider combat is also not real war.

In fact, if we want to identify the moment when individual battles ceased to be characterized by mutual reciprocity, we must look at a time long before artillery, machine guns, and airplanes. Hanson writes extensively on the Greek phalanx. The infantry battle in which the Greek city-states participated was like a collective duel. There was an agreed-on set of both rules and tactics. The two sides decided in advance that the spoils of war would go to the victor—not the win-ner of some total war, but the winner of the battle that would be waged here and now by these men with these lances and shields, on this field, and on this day. "The old style of hoplite conflict," Hanson writes, "was by deliberate design somewhat artificial [and] intended to focus concentrated brutality upon the few in order to spare the many. . . . Success or defeat depended only upon the fighters' ability to stand upright in bronze armor for the next hour or so."[71]

It was not until later, Hanson argues, that military organizations "sought a greater complexity and 'science' to warfare." To ensure the egalitarian, duel-like character of war in the early days of the phalanx, "the javelin thrower, archer, slinger, and their accompanying missile weapons," that is, remote weapons of the seventh century BC, "were kept out of the bloodletting." If battles must be like duels if they are to constitute warfare, then surely twenty-first-century remote warfare operations are not warfare. But neither are artillery, close air support, air interdiction, or even cover-fire operations. War might be like a duel in principle, as Clausewitz suggests. But individual battles ceased to be like a duel when the Greeks abandoned the phalanx after the seventh century BC. To argue that remote warfare has changed the nature of war is to argue that its nature has been changing almost since it began.

The distance that contemporary remote weapons have allowed between warfighters and their weapons' effects is morally significant but can easily be overstated. One of the most important implications of this distance has been the reduction in risk to the people who employ the weapons. But as I have endeavored to show in this chapter, this reduction in risk entails neither a decisive challenge to the warrior ethos nor a change in the nature of war. Neither the war nor the warrior ethos ought to be defined solely by the experience of its participants—experience that continues to change with each sequential generation. Instead, war and the warrior ethos ought to be grounded in more fundamental and unchangeable principles, namely, in the ethics of war. War ought to be defined as a collective activity, and the warrior ethos ought to be defined with reference to the character traits that enable participants to engage in war well. We ought to strive not for the traditional warrior ethos but for the just warrior ethos. The warrior's role, therefore, is not primarily defined in terms of risk to self. Instead, the role is defined properly in terms of the justified harms the warrior causes in defense of others. In Coker's words, "self-respect is won by serving others; it is not self-referential, quite the reverse. It is at the very heart of that much misused term—the 'warrior's honour.'"[72] In many circumstances throughout history, that work has been accompanied by risk to self. But the risk to self is secondary to what it means to be a warrior.

The hellishness of a just war is a necessary condition for bringing about the better peace. The warfighter is indispensable to achieving the end, but the person does so at great moral risk. The task of the warfighter who wishes to be a warrior is to walk the delicate tightrope between justice and mercy—between wielding death and respecting life. Combatants in war—regardless of the physical threat they might face—are invariably faced with a threat to their humanity. This phenomenon is not new. Augustine warned of war's dangers in the fifth century. "Let every one, then, who thinks with pain on all these great evils [of war], so horrible, so ruthless, acknowledge that this is misery. And if any one either endures or thinks of them without mental pain, this is a more miserable plight still, for he thinks himself happy because he has lost human feeling."[73]

The warrior ethos consists of the character traits that instrumentally enable the warrior to walk the treacherous tightrope without falling to either extreme. For the traditional warrior in the Greek phalanx, in the Roman legion, or on the beaches of Normandy, the relevant warrior ethos must include the indispensable requirement to advance in the face of great risk. For the Reaper crew (and a host of other combatants in the modern era), it might not. But this much the Predator pilot and the Greek soldier have in common: Both must cultivate the ability to take life without giving up humanity; to kill but to remain whole; to be willing, if required, to leave their limbs, or even their lives, on the battlefield, but not their respect for human dignity. Philosopher and military ethicist Shannon French has put it this way: "A mother and father may be willing to give their beloved son or daughter's *life* for their country or cause, but I doubt they would be as willing to sacrifice their child's soul."[74] This danger is one that has faced combatants from the beginning.

We can deviate too far in the other direction as well. The love of, and respect for, human life can render a person unable to take it. This too is a risk the warfighter faces. When the warfighter goes too far in one direction, war becomes merely a justification for killing. When a warfighter deviates too far in the other direction, the person refuses to kill at any cost, even when morality demands it. The just warrior maintains the delicate balance—one that requires constant attention and occasional repositioning—an exhausting balancing act between these two extremes. Instead of honor and glory, the warrior pursues justice and mercy in the face of impossible circumstances. When honor and glory do accompany great deeds on the battlefield, it is not because war is beautiful but is because we have caught a glimpse of human virtue shining through the mire.

# 3

# THE MORALITY AND PSYCHOLOGY OF REMOTE WARFARE

I N EARLY 2015, A REAPER CREW TRACKED AN AL QAEDA COM-
mander in the hills of northeast Afghanistan. He was known as
a high-value target —a leader who played a key role in the enemy's
operations. He walked with a young boy, presumably his son. The
Reaper crew waited for an opportunity to strike the target and achieve
mission objectives without causing any civilian casualties, more specifically, without harming the child. Eventually, the two parted ways.
The child walked home, and the al Qaeda leader proceeded to walk
alone away from the house. The pilot and sensor operator took the
shot and killed the man without harming the child. The crew did
exactly what they were asked to do and performed their duties well.
The strike met the demands of *jus in bello* (justice in war)—it was

discriminate, proportionate, and the least harmful means of achieving a justified military objective. In the language of rights, the al Qaeda commander had given up his right not to be killed.

The crew stayed on the scene to conduct the battle damage assessment. The Hellfire missile had not only ended the man's life but also mutilated his body. His limbs lay scattered around the scene. The man's son—a boy who had just lost his father in an attack that was as gruesome as it was sudden—returned to his father's body. Slowly and methodically, he began to pick up the pieces and put them back together in the shape of his late father. I cannot imagine how painful that experience must have been for the child. The Reaper pilot who had taken the shot and who stayed for the battle damage assessment was forced to imagine it. He was a father with a son about the same age as the boy on the screen. He said, "I can't watch this," and asked another pilot to take the controls and left the cockpit.[1]

The psychology of remote warfare has gained considerable attention since the early 2010s. One concern about the ethics of remote warfare arose nearly from the beginning: if war is mediated over thousands of miles through fiber-optic cables and computer monitors, then crews would not experience war as war, and as a result, they would be unable to see the humanity in the people under the crosshairs. The widely held, and reasonable, belief was that if military practitioners were emotionally detached from the death and destruction they caused, the inevitable result would be more death and destruction. In the first decade of Predator and Reaper operations, there was no empirical psychological work with remote warfare crews, and there were so few firsthand accounts available that concerns such as these were common but untested.

Since then, several important studies, conducted primarily by psychologists at the US Air Force's own 711th Human Factors Wing, have provided an empirical basis for a discussion of the psychology of remote warfare. Much of the psychology literature, as well as the ethical discussions that surround it, focus on the negative psychological effects crews can suffer for causing violence remotely. For instance, a Reaper pilot might suffer moral injury or, in the rare and extreme case, post-traumatic stress, for conducting strikes on the other side of the world. Ultimately, though, what has been lacking is an ethical analysis of the positive emotional effects crew members might experience. I introduce these questions at the end of this chapter and provide a possible explanation for them in the next. Ultimately, I hope to explain the relationship between the psychology and the morality of remote warfare. But first, let's look at the two common views that have grown up around remote warfare.

## TWO NARRATIVES

Two competing narratives have dominated the psychology of remote warfare. Early on, one narrative proposed that remote warfare crews are mentally and emotionally disconnected from their operations—that crews had a PlayStation mentality. This idea was later overtaken by the PTSD narrative, suggesting that acute mental trauma is widespread among remote warfare crews. Both accounts miss the mark, with recent evidence providing a more complicated explanation of the psychology of remote warfare.

In 2010, Philip Alston, then the UN special rapporteur on extrajudicial killings, published an influential report in which he coined the term *PlayStation mentality* to describe the supposed psychological disconnection between remote crews and their weapons' effects.[2] The resulting narrative posited that Predator and Reaper crews must treat their lethal work as if it were disconnected from the real world. We were told that killing from thousands of miles away is "like a video game; it's like Call of Duty."[3] We were warned of the "nightmarish image of an 18-year old drone operator basically playing video games from the detached safety of a Nevada bunker."[4] Others said that the operator cannot empirically distinguish between "what he does in obliterating a target and what he does in playing a video game."[5] Predator and Reaper crews were referred to as "cubicle warriors," "desktop warriors," and "armchair soldiers."[6]

The assumption seemed reasonable to many at that time. The reigning theory among many military practitioners was offered by Dave Grossman in his 1995 book, *On Killing.*[7] The now retired US Army lieutenant colonel and psychologist argued that there is a natural human aversion to killing other humans but that this aversion can be overcome with distance. The farther one is from one's target, the easier it is to squeeze the trigger. Because Grossman's work predated the armed Predator, his spectrum of close- to long-range combat went from hand-to-hand combat on one end to artillery and long-range bombers on the other.[8]

Grossman's claim that distance makes killing easier in fact reflects a much older and widely held view. J. Glenn Gray, veteran of the Second World War, wrote some three decades before Grossman: "The man who kills from a distance and without consciousness of the consequences of his deeds feels no need to answer to anyone or to himself."[9] And as we saw in Chapter 2, Clausewitz, writing in the early nineteenth century, recognized that increased physical distance results in a decrease in the "feelings" of war and the "instinct for fighting."[10]

At first glance, remote warfare operations of the twenty-first century seem a natural extension of Grossman's model. If it is psychologically easier to kill using artillery or from a bomber aircraft at twenty thousand feet, how much easier it is from seven thousand miles? An array of theorists deployed this argument to support the PlayStation mentality narrative.[11] The Predator and Reaper represent the ultimate distance in war. Crews in Nevada can kill combatants in Afghanistan. So, shouldn't this kind of killing be easy? Many commentators held that it must be. As late as 2015, Grégoire Chamayou concluded that "the media buzz around the suffering of drone operators was without foundation." Military psychologists, he said, "discovered no trace of post-traumatic stress disorder."[12] The empirical evidence ultimately proved otherwise.

Even before empirical psychology studies became available, the pendulum of public opinion began to swing away from the

PlayStation mentality narrative and toward the PTSD supposition. I once hosted a number of Reaper pilots for a career day at the US Air Force Academy. The nearly one thousand cadets who would graduate the following year could float from table to table and speak to officer representatives from across the air force. Inevitably, those representatives of the various career fields would answer the same questions about a thousand times. During a brief lull, one Reaper pilot leaned back to me and whispered, "Why does everyone want to ask me about PTSD?" The narrative was so compelling that it had seeped even into our own service academy.

In 2012 Brandon Bryant publicly described his own experience with PTSD. Bryant was an air force airman and a Predator sensor operator.[13] He has been the subject of numerous profiles and documentaries, though perhaps the most well known was a 2013 interview with *GQ* magazine. Bryant's descriptions of just how emotionally connected he had been to his work was shocking in large part because it came to light against a backdrop of the PlayStation mentality narrative. In one event that he describes in detail, he and his pilot fired a Hellfire missile at a building that housed a high-value target. In the final moments of the missile's time of flight, Bryant saw something: "This figure runs around the corner. . . . And it looked like a little kid to me. Like a little human person."[14] The intelligence analyst reviewed the video and said it was a dog. The pilot, responsible for writing the after-action report, agreed with the analyst. "The report made no mention of a dog or any other living thing."

Bryant left the air force in 2011, but his experiences left psychological scars. In one poignant moment, he recounts being approached by a teenager who had caught a glimpse of his military ID. The teenager went on to describe his US Marine Corps brother who had "killed like thirty-six dudes." Bryant pounced and told him that if the boy ever talked that way again, Bryant would stab him. "Don't ever disrespect people's deaths like that ever again."[15]

Bryant's public statements have caused contention within the air force's Reaper (and formerly Predator) community. Some have questioned the veracity of his claims of psychological trauma. But at least some of those who flew with him at the time readily admit that the air force was caught off guard by Bryant's self-assessment and was ill prepared to provide him with the medical care he needed. I asked Colonel Travis Norton about Bryant. Norton is a pilot who had then flown alongside Bryant and who has stayed on with the Predator and Reaper community through various levels of command. He acknowledged that the air force hadn't come to grips with the real psychological effects of remote warfare at that time: "[Bryant] came forward and said, 'I'm not well. Something is wrong.' And we didn't take good care of him. The Air Force basically told him to shut up and go back to work. We failed him."[16]

Since that first interview, Bryant has spoken with numerous other outlets, has been the subject of many interviews and at least one documentary film, and has testified before the United Nations. Other sensor operators have since come forward with similar accounts, and in 2015 Bryant, along with three other former airmen, published an open letter to President Obama referring to, among other things, their PTSD diagnoses.[17] As the pendulum then swung too far in the other direction, the PlayStation mentality narrative was replaced by that of PTSD.

During this time, the focus on PTSD was so significant that academics, journalists, artists, and others began to perpetuate the view that acute psychological trauma was rampant throughout the air force's Predator and Reaper community. One early depiction in the arts was George Brant's play *Grounded*, which debuted in 2012 but which was made famous when Anne Hathaway starred in an off-Broadway production of the drama in 2015.[18] The next high-profile debut was Andrew Niccol's 2014 film, *Good Kill*, starring Ethan Hawke as another fighter pilot turned Reaper pilot. The lead characters in both stories suffer from acute psychological trauma that eventually leads them to illegal uses of their remotely piloted aircraft. The 2018 Netflix series *Jack Ryan*, starring and produced by John Krasinski, returned to the PTSD trope with an unhealthy Reaper pilot named Lieutenant Victor Polizzi who engages in a host of self-destructive behaviors. There are countless other examples, and in nearly every case, "drone" warfare is depicted as something so acutely traumatic that few of the crew members involved will escape psychologically unscathed.

Gavin Hood's 2015 *Eye in the Sky* was a welcome departure from the norm in that it depicted the moral stress the Reaper crew might face without assuming acute trauma. The central tension in the film is an operation in which known terrorists are about to conduct a suicide bombing attack against a market—an assault that will undoubtedly kill hundreds. But if the US Reaper crew attacks the terrorists, the Americans will probably kill a child who is in the vicinity. As the story plays out, there are certainly some far-fetched elements. The ability for analysts and operators to know what's happening inside the building—in great detail with both audio and video—rings false. The film also invites viewers to evaluate the most senior decision makers' actions—ministers and cabinet members of the UK and US governments. This also comes across as artificial. Even so, the central question in the film requires decision makers to weigh the loss of innocent life against the consequences of failing to stop terrorists from committing an even greater attack on innocent people. This quandary is a real issue that arises in remote warfare—and indeed in more traditional warfare, too.

Once separated from the military, Bryant was eventually diagnosed with PTSD. He was by no means the only member of a Predator or Reaper crew with this diagnosis, but PTSD is not nearly as common as the Hollywood accounts would have us all believe. Understanding this disorder as well as other psychological effects of remote warfare requires a closer look at the available psychology literature.

One reason that it took the US Air Force and others so long to recognize that PTSD was possible among remote warfighters is that much of the psychological investigation of post-traumatic stress, especially among combat veterans, grew out of the post–Vietnam War era. In what might now be construed as a traditional war, those combatants were physically present in the battlespace, and the acts of violence they committed were accompanied by the physical threats they faced. Many of the Vietnam veterans' cases of PTSD were in fact grounded in the threats faced. But researchers ultimately defined the disorder in overly narrow terms, including risk to self but excluding any other causes.[19] Even now, the fifth edition of the *Diagnostics and Statistical Manual of Mental Disorders* (DSM-5), current as of 2021, suggests that PTSD symptoms will arise "following exposure to one or more traumatic events." The manual constrains the set of relevant traumatic events to "actual or threatened death, serious injury, or sexual violence."[20] As psychologists have further investigated PTSD, however, they have learned that the disorder can result from traumatic events that do not include risk to self.

More recent empirical studies have found that veterans (and others) can experience PTSD not just from being the victim of harm but also from causing harm to others.[21] Some researchers investigated the relationship between PTSD symptoms and remote warfare. The most recent study found that 6.1 percent of remotely piloted aircraft crews who participated in the study meet the DSM-5 symptom criteria for PTSD.[22] This percentage falls within, but at the lower end of, the range of veterans who saw more traditional combat and who experienced PTSD on their return from deployment (an estimated 4 to 18 percent).[23]

The empirical finding is not sensational; only a few remotely piloted aircraft crews experience PTSD symptoms. The disorder is not pervasive across the force, nor has it crippled combat operations. Nevertheless, the finding has profound implications for empirical psychology. As recently as 2010, the consensus among psychology professionals was that a life-threatening trauma or a perceived physical threat to self was a necessary condition for PTSD. In just a decade, the empirical record showed that this condition was not the only cause of PTSD. The community of psychology researchers and the US military made this discovery and instituted responses to it in a remarkably short time.

The history of aviation, and especially military aviation, has often revealed previously unknown facts about human psychology and demanded new approaches to psychological care. The earliest applications of military aviation revealed salient features of human psychology, even if psychologists were not yet fully prepared to account for it. During the First World War, pilots were faced with the dual stressors of flight and combat. The physiological threats of g-forces, hypothermia, exposure in open-air cockpits, and hypoxia due to loss of oxygen at high altitudes arose alongside the psychological threats that would soon be called *shell shock* and, eventually, *post-traumatic stress*.[24] Medical professionals worked diligently to account for, and to provide care for, these newly developing psychological effects. As a British Royal Naval Air Service medical officer named H. Graeme Anderson put it in 1919,

> Aviation, within the last few years, has undergone such enormous developments in the design and construction of machines, making for increased power, stability, speed and climb, that one might be tempted to think that the human machine—the aviator—had been somewhat overlooked. This has not been the case; on the contrary, to some of us in our profession the choice and care of the aviator have proved a new but interesting subject for investigation.[25]

Indeed, in 1917, when the US Army Signal Corps created its novel Aviation Section, it also created the Medical Research Board responsible for investigating pilot health. The board included six subdisciplines, one of which was psychiatry.[26] Yet despite the best efforts of medical professionals, the science lagged behind the lived experience of combat aviators. From the dawn of combat aviation, aviators complained of psychological stress. The medical establishment was, however, slow to recognize that the stress might be psychological reactions to the stressors aviators were experiencing in combat. Medical practitioners tended to attribute negative psychological effects to something like weakness of will, poor temperament, or a deficient

character.[27] When the psychological effects were recognized as the result of combat experiences, medical practitioners lacked the nuanced
categories to describe them—the categories would not be developed
until decades later. In a 1918 report from the British Royal Air Force,
a pilot who suffered negative psychological effects might be described
as having "lost some of his dash" or having gone "hopelessly stale."[28]
In his 1919 book, Anderson describes six categories of negative psychological response to flying, only one of which corresponds to the
stress of combat.[29]

> The writer is convinced that the vast majority of all the cases of "break-
> down" with respect to flying . . . start purely mentally, from an impres-
> sion, an experience, or an act, etc., and that the symptoms and signs
> found later are secondary to the primary mental cause.[30]

By the Second World War, both the US Army's Eighth Air Force
and the British Royal Air Force's Bomber Command were prepared
to respond to psychological effects. The Americans and the British
staffed their various command echelons with psychologists alongside
other medical professionals. In this war, too, airmen would have to
learn to operate in a new kind of aerial warfare.[31] Strategic bombard-
ment, a new way of war, raised new medical challenges. One historian describes some of the dilemmas:

> On the one hand, [military medical professionals] were trained healers
> who tried to get as close to the men as a chaplain would to his flock. . . .
> On the other hand, they were military officers whose principal duty was
> to keep men healthy and sane enough to kill for their country. . . . The
> surgeon's duty was to return as many of those men as possible to the very
> scenes of terror and suffering that had incapacitated them.[32]

Though the psychological diagnoses available during the Second
World War were more nuanced than in the previous war, a wide seg-
ment of the flying force was still deemed psychologically unfit. The

medical establishment didn't make this determination because medical professionals thought the psychological effects of combat had debilitated the flyers but because they found that the aviators lacked a sufficiently strong constitution to stand up to the psychological pressures:

> Senior RAF [Royal Air Force] commanders began to apply an official designation called a "Lack of Moral Fibre" (LMF) to those who refused to fly without measurable physical ailment. The air force shamed those judged to have exhibited LMF by stripping them of rank and privileges, but did not treat them for mental illness or traumatic stress.[33]

Other terms that began to arise during the Second World War included *war weariness, flying fatigue,* and *operational fatigue.*[34] These terms strangely foreshadow the diagnoses remotely piloted aircraft crews would receive more than a half century later, when researchers referred to *high levels of exhaustion, operational burnout,* and *clinical distress.*[35]

If the Predator and, later, the Reaper represented a new kind of warfare, it should be unsurprising that they generated a new set of psychological phenomena. Just as medical professionals had to adapt their profession to meet the needs of a new kind of warfighter in the 1910s and 1940s, so must their counterparts today grapple with the revelation that a person can suffer from PTSD without being exposed to physical risk. Because PTSD had been causally attributed to threat, no one knew that remote warfighters could have this disorder—until some of them did.

The point here is not that the psychological stresses faced by aircrew in the two world wars were the same as those faced by remotely piloted aircraft crews today. But as military technological developments change the way the warfighter experiences war and as psychological professionals work with veterans, they will discover unintended, and previously unpredictable, psychological effects among today's warfighters. Following this same model, the advent of armed remotely piloted aircraft at the beginning of this century likewise changed how these warfighters would experience war and therefore their psychological responses to it. While the air force had been ill-prepared to recognize and treat the trauma experienced by Brandon Bryant and others, it has, along with the medical community, worked through those growing pains and is now in a far better position to recognize and treat psychological trauma among remote warfighters.

These recent psychological discoveries leave us with a dissonance between physical distance and psychological distance. Peter Lee, who has worked closely with British Reaper crews, calls this "the distance paradox." He writes, "Aircraft crews had never been so geographically far away from their targets, yet they witnessed and experienced events on the ground in great detail. In addition, those events were juxtaposed with the banalities of day-to-day family life."[36] US Air Force Colonel Joseph Campo, in his psychological study of remotely piloted aircraft crews, makes a similar point: "The biggest issue society failed to comprehend was the ability for technology to both *separate* and *connect* the warrior to the fight." He adds that distance "is not solely built upon the simple measurement of meters, kilometers, or even continents between combatants. A fuller accounting of distance in warfare accounts for both physical and emotional distance between attacker and target."[37]

One former commander of the 432d Wing that is responsible for the preponderance of US Predator and Reaper aircraft did not couch psychological distance in terms of the distance from pilot to target but described it as the distance from the crew members' eyes to the video monitor that displays the imagery from the aircraft's cameras. "You're 8,000 miles away. What's the big deal? But it's not 8,000 miles away. It's 18 inches away. . . . We're closer in a majority of ways than we've ever been as a service. . . . There's no detachment."[38] Even Chamayou, who in 2015 rejected the claim that remote crews could experience psychological trauma, recognizes this dissonance. "This new combination of physical distance and ocular proximity gives the lie to the classic law of distance: the great distance no longer renders the violence more abstract or more impersonal but, on the contrary, makes it more graphic, more personalized."[39] Remote warfare has challenged the long-standing relationship between distance and the psychology of war.

This complex dissonance between physical distance and psychological distance was also foreshadowed in previous wars. For example, though early in Gray's book, he describes killing from a distance in cold and disconnected terms, by the book's end he makes a much more nuanced claim. "Evidently there is little or no relationship between physical and psychical nearness, for it is possible to be alienated from one's own roommate and be near to someone a thousand miles removed in space."[40] The sense of being close to and yet far from combat is not unique to twenty-first-century remote warfare crews. Perhaps the closest corollary is found in the Allied aircrew who were stationed in the United Kingdom during the Second World War. Miller describes the discontinuity: "In this incredible war, a boy of nineteen or twenty could be fighting for his life over Berlin at eleven o'clock in the morning and be at a London hotel with the date of his dreams at nine that evening."[41] The Reaper crew members are not fighting for their lives, but they are experiencing a frequent transition between far-off war and relative peace at home. One obvious difference between the airmen of the Second World War and today's Reaper crews is that the Allied airmen's psychological transition from war to peace was made possible by a geographic transition from the skies over occupied Europe to London. For remote warfare crews, this psychological transition from war to peace occurs without a change in geography.

This perhaps-surprising revelation that Predator and Reaper crews are both near to and far from their targets validates the proposal in this chapter, namely, that a one-dimensional concept of distance as merely physical is insufficient to describe remote warfare. Even as psychologists and commanders have come to recognize this dissonance, there is work yet to be done. Though the percentage of crews with PTSD symptoms remains low, there are a host of psychological responses—both positive and negative—that fall well short of the acute symptoms of PTSD. If only 6.1 percent of crew members suffer from PTSD, how many suffer from moral injury? I turn to that open question in the next section.

## MORAL INJURY

There is, of course, more to the psychology of killing than PTSD. Jonathan Shay, a psychologist with extensive experience working with Vietnam veterans in the VA (US Department of Veterans Affairs) system, coined the term *moral injury* in his 1994 book, *Achilles in Vietnam*.[42] Since then, and especially since veterans of this century's wars in Iraq and Afghanistan began returning home, moral injury has received more attention among psychology professionals. Moral injury is not incorporated into the DSM-5. While there is no single, universally agreed-on definition, one commonly accepted explanation is that someone can suffer moral injury if he or she witnesses or participates in an event that transgresses his or her deeply held beliefs about humanity.[43]

Because moral injury is such a broad category, it is not obvious which acts should be considered transgressive.[44] When Shay introduced the term, he had a narrow scope in mind. Most of the cases he describes involve a soldier who feels betrayed either by a senior officer or by the army as an institution. One limitation of this view is that it locates the moral agency in the person or group of people who do the betraying.[45] But what if the military member is the one who has committed the moral failure? As the empirical psychology surrounding moral injury has matured, psychologists have broadened the scope to include actions for which a person is morally responsible. To paraphrase one study, combat veterans can experience the effects of moral injury because combat is one of few situations in which a person can experience trauma not just as a victim of violence but also as a perpetrator of it.[46] By participating in actions that transgress their deeply held beliefs about humanity, people expose themselves to moral injury.

Just because moral injury is separate from, and often lacks the acute trauma of, PTSD, it is no small thing. Long before the term *moral injury* was coined, Gray described it—and the way a soldier might respond to it—in combat:

> It is a crucial moment in a soldier's life when he is ordered to perform a deed that he finds completely at variance with his own notions of right and good. Probably for the first time, he discovers that an act someone else thinks to be necessary is for him criminal. His whole being rouses itself in protest, and he may well be forced to choose in this moment of awareness of his freedom an act involving his own life or death. He feels himself caught in a situation that he is powerless to change yet cannot himself be part of. The past cannot be undone and the present is inescapable. His only choice is to alter himself, since all external features are unchangeable.[47]

Moral injury, of course, falls on a spectrum. Perhaps not all examples will be as life-changing as those Gray describes. He seems to suggest here that a soldier must "alter himself" only if the task he is required to perform is, in some sense, beyond the pale—so far outside what he believes to be morally justifiable that he knows that committing the act will injure him. And moral injury is often characterized by feelings of guilt and shame.[48] The counterintuitive revelation from a number of recent studies, however, is that combatants can experience moral injury even when the acts they commit are morally justified and even when they commit no wrong.[49] This finding is a surprising phenomenon. Taking a human life can be a traumatic thing, even when doing so is morally justified, as it often is in war.

As military commanders became aware of the peculiar stressors on remote warfighters, they created positions for operational psychologists, medical doctors, and physiologists in addition to the chaplains who have long been a part of combat organizations. At Creech Air Force Base, some of these positions have coalesced to form what is now called the Human Performance Team. One of the most important features of the team is that the members have the same security clearances as the crews conducting the missions.[50] Rather than waiting for a crew member to make an appointment at the psychologist's office, for example, the psychologist can now walk into the various squadrons and classified work centers and build relationships with people where they are more comfortable.

As members of the Human Performance Team attest, PTSD is just the narrow and acute tip of a much broader psychological iceberg. There are normal and natural psychological responses that fall well short of the PTSD threshold. Creech's wing chaplain, Major Joel, told me, "I'm seeing some moral injury to different degrees."[51] Chaplain Joel described moral injury in terms of the "contemplative nature" of what happens after a pilot pulls the trigger and the sensor operator guides the weapon to the target. "[They] pull the trigger and see that somebody's dead as a result. . . . There's a processing— emotional, spiritual, ethical, mental processing that happens. . . . And so, I've seen some of that across the board." He said most of the moral injury he has encountered among Reaper crews has resulted from the inadvertent killing of civilians. The chaplain suggests that mistakenly killing civilians can amount to a violation of a crew member's values system, "which is textbook moral injury."

In other cases, though, the person killed was a lawful target and no one else was physically harmed, but the targeted individual's family members either saw the strike or came to collect the body. This was the situation described at the beginning of this chapter. Not only did the pilot see that he had taken a combatant's life, but he also saw the moral cost borne by the child whom that father left behind. In the chaplain's words, these are cases in which "women and children—family members—[are] witnessing the death of the bad guy." Crew members recognize that even though the person killed was a morally and legally viable target, "he's still a father. He's still a husband. [Some crew members respond by saying] 'I'm a father. I'm a husband' and reacting because they see the family's reaction. . . . They're processing it at a very personal level."

Despite Joel's confidence in the value of the Human Performance Team, he thinks that the first line of defense against moral injury ought to be a person's fellow crew members. Even though Joel has all the relevant clearances—he can sit with crews in the cockpit and review strike videos with them after the fact—he's a chaplain and a noncombatant by law.[52] He says, "The front line of defense is someone who's got experience. . . . I've never pulled a trigger, and I never will. [Crews need to] talk to somebody who's got experience doing this."

Master Sergeant Sean, an instructor sensor operator, told me about a time he did just that. A young airman was tasked with supporting friendly forces on the ground. The airman looked on as one of the soldiers he was supporting stepped on an improvised explosive device (IED). The crew watched, helpless, as the explosion killed one of the soldiers they were responsible for protecting. There was "nothing he could do about it," Sean said. "I sat down with that airman and talked with him for like two hours, just to make sure that his head space was really good and where he needed to be, and also that there were opportunities to go talk to mental health and the wing psychologist."[53]

Like the military medical professionals in the Second World War, Creech's Human Performance Team is responsible for caring for airmen, in part, to ensure they are mentally prepared to perform their professional duties. Sometimes they are unsuccessful. Dr. Richard, the wing psychologist, told me that if someone experiences a devastating event, his or her mental representation of the event is malleable for just a short time and providing care during that period can be crucial. Dr. Richard told me about one case in which an airman refused to continue in the job and had to be retrained in a different air force career field. "I can't do this anymore," the airman had told Dr. Richard. "I don't care what we say in here. We can talk. You can tell me what you think, but . . . I'm not doing this anymore."[54]

In the aforementioned Campo study, participants among remote warfare crews self-reported both positive and negative psychological responses to killing. I will return to the positive responses later in this chapter. Many of the negative responses fell far short of the acute standard required for a PTSD diagnosis. Yet the negative responses are undeniably real. One participant recounted a story in which everything went right, that is, the missile struck the intended target, all those involved had every reason to believe that the attack was justified, and there were no civilian casualties or other collateral damage. Nevertheless, the strike had a profound impact on the crew member.

> We kill him . . . that's the first time I saw someone dead and we zoom in to view the dead body and get BDA [battle damage assessment]. Right then, it hit me. My heart just started pumping. I went home that night and couldn't talk with my wife. She knew something was wrong. I couldn't get that image of his [dead] body out of my mind. Then about four days later I started thinking about a kid growing up without his father that I had killed. The humane thing is to let him live, but this guy was trying to kill Americans. Finally, about two weeks later I broke down. I couldn't hold it in anymore and I had to seek help. . . . I wanted to know if God was OK with what I was doing.[55]

Nothing in this crew member's account suggests that he is reliving the traumatic event, remaining hypervigilant, or experiencing any of PTSD's other well-known symptoms. But he has experienced a significant negative psychological effect. While the narratives may differ in the details, according to Campo's study, 33 percent of crew members experience negative psychological effects to some degree or other following their first employment of lethal force.

Unlike questions about post-traumatic stress, no available empirical record tells us whether or to what degree remote warfighters experience moral injury. We lack this information in part because, as a relatively new field of study, psychology researchers have yet to settle on standardized definitions, symptoms, and metrics for measuring moral injury. Meanwhile, the studies that have involved remotely piloted aircraft crews have also overlooked moral injury.[56] Until empirical psychology catches up to these developments in the lived experience of war, and until those sharpened analytic tools are applied to remote warfare, we cannot be sure about how significant moral injury is among remote warfighters.

The effect the Predator and Reaper have had on the psychology of war might extend beyond remote pilots and sensor operators. Not only do technological developments after 2001 bring the pilot and crew empathetically closer to the killing, but they also bring the whole warfighting enterprise closer. The same video of the surveillance leading up to a strike, the strike, and the aftermath that is seen by the pilot and sensor operator is often seen by almost everyone involved in the mission. Remote-operated video-enhanced receiver technology allows ground units, as well as other aircraft, to view the full-motion video feed.[57] The Distributed Common Ground Station—a network of networks that pushes the feed around the world—publishes the real-time video to tactical operations centers, the air operations center, and the numerous intelligence agencies, organizations, and operations centers that support the remotely piloted aircraft mission.[58] The project has captured the deadly realities of war from the front lines and distributed them across the vast array of operations, support, and contract personnel.

One air force intelligence professional, Lieutenant Colonel Matthew Atkins, writes in an essay, "The nature of warfare actually has shifted [the psychological] burden to the *finders* of targets, the intelligence and special operations personnel that identify the people that need to die." He concludes,

> It is clear that our ongoing counterterrorism missions are not cold, sterile operations that make it easy to kill. On the contrary, our [remotely piloted aircraft] and intelligence personnel are engaged in an intensely personal hunt where they can count the children of the terrorist they seek to target. The human in the loop always makes the decision. And the

human in the loop always bears the consequences of making that life or death decision.[59]

Since Atkins's essay was published in 2014, intelligence professionals have indeed come forward with grievances about the psychological effects of supporting remote warfare operations. Some have also found fault with the care they have received. Sonia Kennebeck's 2017 documentary, *National Bird*, follows three intelligence analysts who supported Predator and Reaper operations either as active-duty airmen or as contractors. At least one of the three, a former analyst named Heather, was eventually diagnosed with PTSD.[60]

As was the case with pilots and sensor operators, the long-term psychological costs to these intelligence professionals, now intimately familiar with the people in their target folders, is still largely unknown. Though current telecommunications technology does make some elements of intelligence analysts' experiences similar to those of the pilots and sensor operators, we must not ignore agency. As the preliminary psychological investigations cited previously have shown, some of the psychological implications of pulling the trigger, releasing the weapon, and guiding it to the target—willfully taking the life of another human being—will not be shared by intelligence personnel who observe these events, no matter how high-resolution the video. But the intelligence professionals, who sometimes have identified the target who will ultimately be killed, might experience unique psychological effects. We must put this thought aside for the moment, however, and give psychological study the requisite time to catch up to rapidly developing military experience.

## CAUSE FOR CELEBRATION?

One important question remains. Morally speaking, what might explain the positive emotional reaction many crew members have expressed after taking the lives of combatants on the other side of the world?

The PTSD narrative that we find in Hollywood films and TV shows assumes that the psychological effects remote warfighters experience are always negative. But the empirical evidence tells a different story. One study found that a majority of Predator and Reaper crew members who release a weapon in combat for the first time experience a positive emotional reaction. About a quarter of them experience both positive and negative emotional responses.[61]

This empirical claim that crews experience both positive and negative emotional responses to taking human life is jarring. How can a person feel good about taking a life? So much of the discussion on the psychology of remote warfare has focused either on emotional detachment, as in the PlayStation mentality narrative, or on trauma that remote crews might suffer, as in the PTSD narrative. The study that points to positive and negative emotional responses suggests a much more complicated picture of the psychology of remote warfare.

The accounts provided in the study suggest that the positive emotional response is generally based on the crew members' sense that they had done a job well, completed a difficult task, or supported friendly forces on the ground.[62] A Reaper pilot named Lieutenant Mister told me about the first time he released a weapon in combat. "After my first strike," he said, "I had a positive emotional response. We were, like, cheering." I asked him to go into more detail—to explain what precisely he was cheering about. He responded,

> I was happy that the sensor [operator] got the crosshairs on the people. . . . So that was my immediate response—that we didn't screw up the shot. . . . The comm[unication]s were right. . . . [I was happy] that we all worked together as a team. . . . And then I was also happy because there were safety observers behind us, and they did their jobs. . . . I was just overall happy with the entire situation.[63]

Notably absent from Mister's description of his celebratory attitude is any reference to the death and destruction he had caused. The moral and psychological landscape Reaper crews occupy is complicated because they—like so many warfighters before them—cause harm to prevent harm. To put it bluntly, they have to kill some people to prevent other people from being killed. This condition is common to many warfighters, but because the crew's lives are not threatened, this relationship is not as obvious in remote warfare.

For millennia, combatants have returned from war offering variations on the theme "It's about the person next to you."[64] In what we often think of as traditional war, the soldier in the foxhole tries to kill the enemy before the enemy can kill him and his friends. In the Reaper case, the defense-of-others theme is extended in time and space but is still present, a point to which I return in the next chapter.

The most experienced pilot with whom I spoke had flown a dozen different aircraft in a wide array of missions before arriving at Creech to fly the Reaper. He said, "[You're not] celebrating the fact that you've taken human lives. [You're] celebrating the fact that you're saving human lives. . . . I've done what I needed to do in a surgical way. I've taken the right life."[65]

Another pilot, a former marine named Lieutenant Steven, had conducted his first strike just days before our meeting. He told me that he had planned to treat himself to a high-calorie dinner to celebrate his first shot. "But then after [the shot], I didn't really feel like eating. I didn't feel sad or angry. It just felt weird. I stayed up that night and thought over and over about what I could have done better."[66] As soon as Steven said this, another pilot said, "I had the exact same experience."

I asked Dr. Richard about the emotional response that crew members can have to a successful strike. He asked me if I had ever participated in competitive sports and used this analogy:

> You remember the amount of effort and time that you put into that, and you achieve that goal, and as soon as you finish, it's like "Boom, I feel great." What's going on? Dopamine is popping off in the brain. Then you get people coming up to you and saying "Great job." . . . Now you got oxytocin popping off in the brain. . . . And then you have serotonin,

which is like pride: "I completed this. I did a great job. Everything . . . came to fulfillment." [The crews] aren't glorifying the kill. [They're] getting this physiological response. It's going to happen. It's human nature. . . . We [tell them], "Yes. Feel every ounce of appreciation and pride in what you do; you spent a lot of time [training]. . . . This is honorable work."[67]

Air force physiologist Major Maria put it a little differently. She said there's a difference between taking pride in a job well done and taking joy in a violent act.[68]

The moral, emotional, and, for some, spiritual demands placed on the Reaper crews are complicated. The crews are asked to strive for technical and tactical excellence in their work. And when their work involves justifiably taking human life, they are asked not to let it affect them. In the rare cases of tragedy—when civilians or friendly forces are inadvertently killed—the crew members are asked to empathize with the people on the other end of their aircraft's cameras: to work through the normal human responses of regret and remorse but not to become unraveled by it. For many, the result of this complexity is a fractured compartmentalization. The professional desire to do well, the natural motivation to protect forces on the ground, the emotional toll of taking human life, and relationships in their home lives are all carefully sterilized, sorted, and stored. But it is not at all clear that the model is sustainable.

Neither the PlayStation nor the PTSD narrative captures the psychological realities of remote warfare. Crews experience psychological effects similar to those their more traditional counterparts experience—with the obvious difference that because they are not personally being attacked, remote crews do not experience the psychological effects of fear and life-threatening trauma. As the field of empirical psychology has matured in recent decades, however, psychology researchers increasingly recognize that to a considerable degree, the trauma and other psychological effects that warfighters have experienced in the past may have resulted from acts of killing as well as the risk of dying.

There are two important conclusions. First, the PlayStation mentality and the PTSD narratives are insufficient to capture the nuances of psychological distance in remote warfare. With some exceptions, the crews are emotionally engaged in their lethal work yet not ravaged by PTSD. They experience psychological trauma at rates similar to, if not slightly lower than, those of more traditional military members. Second, psychological and physical distance are different in important ways. Throughout the history of war, many technological developments that increase the physical distance between shooter and target have increased the warrior's psychological distance, but that's not the case with remotely piloted aircraft. Because crews are presented with visual evidence of the violence caused by enemy forces, can see the violence they themselves cause to the enemy, and can loiter over the target, they respond emotionally and psychologically as if they were much closer.

Although the psychological costs borne by remote warfare crews are not often acute, the burdens are nevertheless real. The pilot whose story opens this chapter will forever live with the image of that boy collecting the pieces of his father. Though the pilot might have had some positive emotional response to doing a job well or protecting the terrorist leader's would-be victims, he bears responsibility and memory too. And he does so in the service of his country. The psychology of remote warfare is complicated.

# 4

# GOOD GUYS AND BAD GUYS

## WHEN IS KILLING JUSTIFIED?

Are remote warfare operations morally justified?

In 2009, Baitullah Mehsud, the leader of the Taliban in Pakistan, was killed while he was undergoing dialysis on his father-in-law's rooftop.

In 2011, the United States killed American citizen Anwar al-Aulaqi while he traversed the Yemeni desert.

In 2020, the Iranian military leader Qasem Soleimani was killed as he departed Baghdad airport.[1]

There was no firefight. There were no US forces being shot at. These men and many others like them were killed out of a clear blue sky by a Predator or Reaper crew half a world away. How can this kind of warfare be morally justified?

One of the most important arguments against remote warfare is that a combatant who faces no combat risk has no moral justification for killing enemy fighters. To answer the question of whether these and other killings have moral justification, we have to look at self-defense, the defense of others, the just war tradition, and the morality of defensive harm.

The moral justification is easy to see in close air support operations. The Reaper crew works not for its own sake but in defense of others, fellow service members on the ground. Other cases, such as a high-value targeting operation, are more difficult. The crew will, for example, target a terrorist or an insurgent leader even while he poses no imminent threat. In these cases, there is no firefight; there are no US service members under attack. Nor is there an imminent threat—at least not in the ordinary use of that term.[2] How can strikes such as these be morally justified? As we will see, high-value targeting operations can be conducted in the defense of others. But some have argued that Predator and Reaper operations do not fall under the same category as other military operations. Even if just war theory permits traditional fighters to shoot at one another in self-defense and in the defense of others, do Reaper crews seven thousand miles from the target area fall into the same category? Do those permissions apply to a combatant who is physically removed from combat?

Though we might think of these aircraft as operating in isolation, disconnected from the broader military, they are often supporting a team of soldiers, marines, or special operators on the ground. Sometimes this connection between a Reaper crew and those they support is clearer than others. In one account, a sensor operator, a technical sergeant named Jesus ("heh-soos"), slewed his sensor ball to a set of coordinates just outside Kandahar Air Field, Afghanistan. As a sensor operator, Jesus is an enlisted airman who sits in the right seat and is responsible for the sensor ball and all its cameras as well as its targeting laser and infrared pointer. While the pilot flies the aircraft, the sensor operator "flies the ball." The pilot is an officer and, in addition to flying the airplane, maintains responsibility for the entire aircraft and crew. While Jesus moved the camera to find the coordinates he was given, the pilot tried to reach the joint terminal attack controller (JTAC, pronounced "jay-tack") on the radio.[3] The JTAC is an airpower professional, usually an airman, embedded with the ground team. Jesus and his pilot were in their ground-based cockpit on Kandahar Air Field, uncharacteristically close to the JTAC and the rest of the ground force. The aircraft returning to the base had excess fuel and didn't need to land right away. In cases like this, the launch and recovery crew can sometimes fly the airplane past its scheduled land time and offer support to the base defense operations center. In this case, the crew called base defense and asked for a tasking before they even stepped to the cockpit. The base defense operations center was responsible for the security of Kandahar Air Field. This sprawling base in the lowlands of southern Afghanistan was then home to more than twenty thousand people from across the coalition.

It was the spring of 2012, and, not far from the base perimeter, a platoon-sized US Army unit had been pinned down by enemy sniper fire. By the time the airplane arrived on station, the shooting had stopped. But three of the soldiers had been wounded and had to be medically evacuated. The JTAC told the crew that the sniper fire had been coming from a nearby tree line. The crew members searched for the sniper, but to no avail. Though the firing had stopped at least for the time being, the soldiers faced another problem. The enemy force that had attacked them would probably also attack the rescue helicopter—jeopardizing both the helicopter crew and the wounded soldiers' means of getting to the hospital. The Reaper crew's primary mission shifted. Rather than finding and engaging the sniper, their task now was to find a safe helicopter landing zone, coordinate with the inbound helicopter crew, and provide overwatch in case the enemy fighters attempted to attack the helicopter. The plan the Reaper crew devised was a good one, and the medical evacuation was successful and relatively uneventful. Without firing a shot, the Reaper crew had established trust with the ground forces and helped ensure that the three wounded soldiers got to the hospital quickly and safely.

That night, Jesus and some of his fellow crew members were given an opportunity that is rare in the Reaper community. The commander of the squadron, a pilot named Lieutenant Colonel Julian, had arranged for some of the Predator and Reaper crews to visit the three wounded soldiers at the hospital on base. Julian told me that his intention was for the Predator and Reaper crews to say thank you to the soldiers. "For me it was about showing our gratitude for their sacrifice and acknowledging that they were in harm's way. And then it was also about connecting our crews to allow them to see their contributions." Julian wanted to give the airmen in his squadron the opportunity to meet these soldiers face-to-face. When the Predator and Reaper crews conduct missions from the United States, he said, "you never truly have that personal relationship with those [soldiers] on the ground. . . . So it was great to see things from their perspective."[4]

Visiting the Kandahar hospital that night had a big impact on Jesus. Reluctant to go at first, he made the visit only because it had been Julian, a commander he respected, who suggested it. His hesitation wasn't because of a lack of interest in the soldiers' recovery. Instead, it was because he had never met someone whom he had supported on the ground. He had always taken his duty to support ground forces seriously and had always done his best to serve them well. But for him, there was something impersonal about supporting friendly forces through fiber-optic cables and video screens, and he wondered if that would change once he had met these soldiers in person. It didn't take long, though, for him to understand why Julian had encouraged the visit. It was, as Julian had anticipated, an opportunity for army soldiers and air force airmen to connect—human to human. Julian had heard that when the soldiers dove for cover in response to the sniper fire, one soldier landed in a drainage ditch full of water, waterlogging and destroying the portable Sony PlayStation in his cargo pants pocket. When members of the squadron arrived, they brought a few PlayStations for the soldiers. One of the wounded, a man named Dakota, had been shot in the shoulder and in the thigh. A third bullet, Jesus told me, had gone "through his helmet. It scraped the back of his head and actually scraped the skin off, that's how close the bullet was." Dakota took off the patch he had been wearing during the sniper attack and gave it to Jesus. "Thank you for doing this," Dakota said. Jesus told me later how the visit affected him:

> Meeting someone on the ground that I've supported felt surreal. We work in an environment where we are normally separated from the fight through [satellite communications]. . . . To actually meet someone that you help protect and them saying to you, "Thank you for looking over us," is one of the best feelings in the world. This is why I enjoy my job as a sensor operator.

He still has Dakota's unit patch. On that day in 2012, the Reaper crew released no weapons, and no enemy combatants were killed. Yet the Reaper crew provided support to a US military unit that needed it. When friendly forces are under attack, the Reaper crew's moral justification is obvious. They're defending ground forces that need to be defended. But not all Reaper sorties are like this.

## THE MORAL JUSTIFICATION FOR KILLING IN WAR

Most people think killing in war can be morally justified, though pacifists are an important exception. Most of us believe that when two political communities go to war, the combatants on the one side are morally justified in trying to kill combatants on the other. The prevailing view is that, for example, when an American soldier shot at an Iraqi soldier in 1991, or when a Russian soldier shot at a German soldier in 1945, or when a Union soldier shot at a Confederate soldier in 1865, each had a moral justification for doing so. We generally don't think that soldiers on one side do something morally wrong when they attack soldiers on the other side. But why not? If you share the perception that killing in war is morally justified, at least under some conditions, then you might think this a strange question. But concerns over whether remote warfare can be morally justified demand that we ask it. As we saw in the two previous chapters, remote warfare separates the act of killing from the risk of dying. Just as the separation of these two concepts raises questions for the warrior ethos and for psychology, it also raises questions about the moral justification for killing in war.

Outside of war, moral justifications for killing are rare. In other words, in ordinary life, there is a strong moral prohibition against killing, and this prohibition is rarely outweighed. I once had my nose broken in a fight in which I was an unwilling participant. By the time he finished, my assailant had punched four or five people in the face. I believe now as I did then that I had a right *not* to be punched in the face. When my assailant punched me, he violated my right. There was little disagreement on this point. My friends, some passersby, and the police all agreed that I had a right not to be punched in the face. In ordinary life, acts of violence are often coincident with rights violations—when we hear that someone has been physically harmed, the person's right not to be harmed has often been violated, too. And of course, if I ordinarily have a moral right not to be punched in the face, then surely I ordinarily have a moral right not to be killed. You have a corresponding moral obligation not to kill me.

These rights and duties often go hand-in-hand. I have a right to my property, and you have a duty not to take it. I have a right to my freedom, and you have a duty not to lock me up. Under some circumstances, I lose my right and you are absolved of your duty—self-defense can be one of these circumstances. But outside of war, we are rarely placed in situations in which the only way to defend a right is to take a life.

The language of rights and duties in war might seem abstract, but it applies in concrete ways. Most just war theorists agree, as does international law, that combatants suddenly regain their right not to be killed in fighting by laying down their arms and waving the white flag of surrender. They make the transition, then, from a legitimate target to an illegitimate one. But according to other provisions in international law, combatants might give up their right not to be killed but retain other rights, for example, the right not to face excessive suffering. The Hague Convention in 1899 banned the use of expanding bullets because they cause unnecessary harm. Glass projectiles are prohibited because they are not discoverable by X-ray and are therefore difficult for medical professionals to find and remove. In both cases, the means is prohibited because it causes more suffering than is necessary to achieve the military objective—it violates the just war principle of necessity.

Assuming that these constraints in the laws of war reflect the morality of war—and I think they do—it seems that combatants quickly lose their right not to be killed even though they retain these myriad other rights. As soon as you don your nation's flag, pick up a weapon, and enter the battlespace, you give up the right not to be killed. The right not to be killed is among the most important rights a person has, but it is among the first to go for combatants. What causes these combatants to give up their right not to be killed?

In our common vernacular, we sometimes talk as though we—the United States and its allies—are the good guys and our enemies are the bad guys. In fact, some of the quotations I use throughout this book refer to "killing the bad guys." On the surface, it might look like this distinction helps explain why enemy combatants lose their rights not to be killed. But what do we mean when we say that this ISIS leader or that al Qaeda fighter is a bad guy? And is whatever we mean sufficient to justify killing him? Perhaps we mean something like, this is a person who does evil things—engages in human trafficking or in systematic rape, as members of ISIS have done.[5] I once heard an RAF general describe ISIS as villains "right out of central casting." If we're looking for a morally reprehensible set of activities, we need look no further than ISIS. But we don't limit our use of the term *bad guys* only to those who conduct atrocities. Instead, we tend to use the term as a shorthand for whomever we are fighting. In the 1950s, the North Korean soldiers were the bad guys. A decade later, it was North Vietnamese Army soldiers and Viet Cong guerillas. And a few decades after that, Iraqi soldiers would be the bad guys.

Many high-value targeting operations are directed against leaders who we can agree are bad guys. On the surface, this bad-guy language seems sufficient to justify the killings of Mehsud, al-Aulaqi, and Soleimani. Mehsud, who cobbled together militant groups to form what became the Taliban in Pakistan, was responsible for dozens of suicide bombings. Al-Aulaqi crossed the line from on-line preacher to operational leader when he planned the attempted Christmas Day airline bombing in 2009. Soleimani has been supporting militia forces throughout the Middle East in their attacks on US forces. He undoubtedly has US service members' blood on his hands. Even so, it seems unreasonable to insist that all the enemy soldiers who were killed in these confrontations were guilty of the kind of morally reprehensible behavior that we can ascribe to many ISIS fighters and terrorist leaders. And, for that matter, we cannot even be sure that every ISIS fighter is guilty of morally reprehensible behavior.

We cannot explain why a combatant on one side of a war has a moral justification for killing combatants on the other side, simply by casting enemy combatants as bad guys. Some enemy combatants are not bad guys, and some bad guys are not enemy combatants. Though the bad-guy term is a common shorthand, it is of no use in determining what causes combatants in general to lose their right not to be killed.

## SELF-DEFENSE

There is another plausible way to describe the moral justification for killing in war. When Dakota and his unit were attacked by the sniper in Afghanistan, did he and his fellow soldiers have a moral justification for shooting back? As the event unfolded, of course, they never did find the sniper—but suppose they had. If Dakota had been able to see, shoot, and kill the enemy sniper, would he have been morally justified in doing so? Dakota would have been acting in self-defense and in defense of his fellow soldiers. Perhaps this is the reason that killing in war is morally justified—because combatants act in self-defense and in the defense of their comrades.

The morality of war, and even the laws of war, are often described in terms of self-defense. A nation's legal right to wage war is codified in Article 51 of the UN Charter: "Nothing in the present Charter shall impair the inherent right of individual or collective self-defense if an armed attack occurs against a Member of the United Nations."[6] If "individual or collective self-defense" is a suitable justification for going to war, then perhaps individual or collective self-defense is a moral justification for specific acts of killing in war.

But this codified justification might pose a moral challenge for remote warfare. If killing in war is justified only by self-defense, then remote warfighters are unjustified in causing harm, because they are not defending themselves. We can recall all the claims about reciprocal risk from Chapter 2. For the ordinary combatant, perhaps the moral justification works something like this: Because they are trying to kill me, I am morally justified in trying to kill them. The criticism many have raised of remote warfare is that these crews do not stand in the same reciprocal relationship. No one was shooting at Jesus and his pilot. What if Jesus had seen, shot, and killed the sniper from his ground-based cockpit? Would he have been morally justified in taking that life? If the moral justification for killing in war is grounded in reciprocal risk, and if remote warfare operations incur no such risk, then do remote crews lack a moral justification for killing in war?

The Kandahar crew members would indeed have been morally justified in killing the sniper—not because of any risk they faced but because Dakota was under attack. The fact that Dakota and his unit were being threatened would have justified the crew's lethal intervention. Thinking of justified violence in war in these terms is not a novel idea. For millennia, warfighters have recognized that often, it's about the person next to you.

Audie Murphy was just a "baby-faced Texas farm boy" when he commanded a US Army infantry company in France during the Second World War.[7] Though he would go on to become a Hollywood movie star, in January 1945, he led his company in its defense against an attack by hundreds of German soldiers and six tanks. After providing cover fire so his soldiers could reposition in a tree line, Murphy called in an artillery strike against the tanks. The company had an M18 Hellcat "tank destroyer," but it had already been fired on by the German tanks and was set ablaze. Murphy nevertheless climbed aboard the flaming wreckage and fired its still-functioning .50-caliber machine gun against the attacking German soldiers. From that position, a wounded Murphy fought on for over an hour before leading his company in a counterattack. Years later, long after Murphy was awarded the Medal of Honor and had established himself as a Hollywood A-lister, he was asked to explain how he could act so courageously under such devastating conditions. Murphy responded simply, "They were killing my friends."[8]

When we picture stereotypical combat—soldiers on one side shooting at soldiers on the other—we imagine each soldier fighting for his or her life. At the same time, there is another sense in which each fights for brothers and sisters in arms. It is not just each soldier's life that is at stake but also the lives of the rest of the soldiers in the unit. When we think about self-defense and other-defense, these are the ideas that often come to mind. And so, when service members come back from war having done terribly violent things, we sense it is right to say, "You did what you had to do. It was kill or be killed." Likewise, this twofold concept helps us understand Audie Murphy's explanation for his courageous exploits: "They were killing my friends." But a soldier's reasons for acting and a soldier's moral justification are not the same thing. When we think about soldiers on a notional battlefield, defense of self and defense of fellow comrades are the most obvious reasons that a soldier might be morally justified in killing enemy soldiers, but there are other reasons more central to the morality of war.

Even before the age of remote warfare, many of us took it for granted that service members doing the fighting might face varying levels of risk. Even among ground forces, we might think of snipers as soldiers who are in a position to defend, say, infantry soldiers, even though the risks they face are not the same. With snipers, there is a voluntary distribution of risk. There is the tacit agreement that snipers will take on less risk than the average infantry soldier will, so that the sniper can be in a better position to provide more effective fire when the infantry soldiers most crucially need it. We see this same voluntary redistribution of risk and force in the most basic small-unit tactics. A small unit might advance on an enemy position in teams of two: the first person runs while the second person shoots. Then they switch roles. At any given time, one person is accepting more risk, while the other is imposing more force.

If we abstract one level from small-unit tactics, we find different army branches supporting one another in the land fight. Imagine that the team on the ground consists of infantry soldiers advancing on an enemy position as well as snipers providing armed overwatch of their positions. Off in the distance, there is an artillery battery prepared to provide indirect fire in support of the ground team's mission. Now, the team leader on the ground can ask those who face substantially less risk than do the infantry soldiers—because they remain some distance from the fight—to provide significantly greater firepower. If so, even though the artillery team might not face any direct threat at the moment, they are justified in causing harm to the enemy to defend their fellow combatants on the assault team.

The picture continues to widen when we add airpower in combined-arms warfare. In addition to the infantry, snipers, and artillery, there are also aircraft that, under the authority of the ground force commander, bring force to bear against enemy combatants. Though in some combat situations, this close air support work can be very dangerous to the aircrew, in the US wars since 2001, aircrew members have generally faced far less risk than their counterparts on the ground have experienced. In these cases, the pilots and crew members face less risk, but they are nevertheless morally justified in committing acts of violence to defend ground forces being threatened by the enemy.

The idea that a combatant can be morally justified for defending others is pervasive in war. It is easy to fit Jesus and Dakota into this frame. When the sniper was shooting at Dakota and the rest of the ground unit, artillery batteries and traditionally piloted aircraft would have been morally justified in killing the sniper. Likewise, the crew, operating from a few miles away on Kandahar Air Field, would also have been morally justified in killing the sniper. In fact, even a Reaper crew seven thousand miles away in Nevada would have been justified in defending Dakota. This moral justification is grounded not in self-defense, but in other-defense. These soldiers and airmen would have been morally justified in defending the ground force that was under attack.

Thinking of remote airpower operations in terms of other-defense raises two important questions. The first concerns the moral constraints on those operations. Even if a combatant is morally justified in causing harm to defend others, how much harm is the person permitted to cause, and under what circumstances? The second question addresses high-value targeting operations for which remote warfare has become well known. What if there is no sniper actively shooting at friendly forces? Can a remote warfare crew be morally justified in causing harm under those conditions? When most people imagine drone strikes, they probably don't think of close air support but are imagining high-value targeting. And this question is central to the misgivings many have about remote warfare. How can it be morally justified to target and kill a person while he poses no immediate threat to friendly forces?

## JUST WAR THEORY

Just war theory provides an account of the morality of war, and though the theory has evolved over time, many scholars agree that its founding contributors were Aristotle (third century BC), Cicero (AD first century), and Augustine (AD fifth century). Broadly speaking, just war theory makes claims about when the decision to go to war is morally justified, which acts within a war are morally justified, and how participants ought to behave after the end of hostilities. These three segments are often referred to respectively by the Latin terms *jus ad bellum*, *jus in bello*, and *jus post bellum*. Because my concern in this book is at the tactical level of war, I am concerned chiefly with *jus in bello*—justice in war. There is some disagreement among just war theorists as to how the *jus in bello* requirements ought to be listed, but most theorists agree that in a military conflict, an action is justified only if it is discriminate, proportionate, and necessary.

A combatant on one side must distinguish—discriminate— between those who are participants in hostilities and those who are not. Moreover, the combatant must target only the former and re- frain from targeting the latter. Proportionality stipulates that, for any military action considered, the anticipated good to be achieved must outweigh the resultant moral costs—though some theorists say what matters to proportionality is not the moral good that results but the anticipated military value. Finally, necessity—sometimes called *min- imal harm*—requires that the course of action be the least harmful means of achieving the justified end.[9]

When theorists began looking at remote warfare operations in the last fifteen years or so, just war theory seemed unable to account for the misgivings many had. There were certainly individual cases that failed to meet one or more of the *jus in bello* standards. Predator and Reaper crews have, for example, made mistakes that have caused un- necessary deaths of noncombatants.

Perhaps the most devastating example occurred in 2010. The *Los Angeles Times* obtained a transcript of the Predator crew's conversa- tion before and after a helicopter attack.[10] This transcript remains one of the only widely published dialogues between members of a Predator or Reaper crew. The crew was convinced that the people un- der its crosshairs were terrorists with weapons, preparing for an early morning attack on US military ground forces. In reality, the people killed were not terrorists, and they had no weapons. They were men, women, and children getting an early start on their journey to Kabul. Having no idea that there was a US Army unit nearby, they had no reason to think US forces would find their behavior suspicious. The account is gut-wrenching, and the event tragic. Those innocent people should never have been killed.

But concern about cases in which crews made mistakes—even tragically killing the wrong people—still does not account for the moral misgivings people have about remote warfare, at least not entirely. Their misgivings are about remote warfare as a practice, or about the Predator and Reaper as specific weapons systems, not just about individual mistakes. After all, other US combatants make mistakes too. In 2014, a B-1 Lancer ("Bone") bomber inadvertently dropped munitions on friendly forces at least partly because the bomber crew failed to recognize the infrared strobes on the Afghan soldiers' helmets in its infrared targeting pod.[11] A year later, an AC-130 gunship fired on a Doctors Without Borders hospital in Kunduz, unaware that it was a hospital.[12] Both cases were tragic, to be sure. But cases like these haven't caused commentators to argue that the B-1 or the AC-130 are unethical as systems, or that employing airpower is unethical as a practice. But many theorists have made that claim about remote warfare. Aside from individual cases in which crews made mistakes, there seemed to be nothing in the nature of remote warfare that violated specific principles of *jus in bello*. Remote warfare operations can be just as discriminate, proportionate, and necessary as their traditionally piloted counterparts.

Imagine that a Reaper crew is seeking to target an enemy combatant in Afghanistan. If the target is actually an enemy combatant and if the strike would harm him alone and cause no harm to noncombatants and no damage to civilian property, then surely the strike is discriminate. If the target is going to kill people—either combatants or local civilians—then one combatant life is being taken to prevent the killing of numerous other people. The strike is proportionate.

Finally, if this strike is the only way—or the least harmful way—to keep this enemy combatant from conducting his harmful attack, then the strike is necessary. The three conditions are met, and the fact that the strike will be carried out remotely seems to pose no significant challenge to the morality of war. To many, this is an unsatisfying conclusion: there is something morally questionable about remote warfare—not just in discrete cases but as a practice. But at first glance, the *jus in bello* principles of discrimination, proportionality, and necessity cannot explain those misgivings.

This conundrum drove some theorists to dive deeper into just war theory. Perhaps the issue was not that Predator and Reaper operations violated a specific just war principle but was instead that just war theory assumes certain preconditions that the Predator and Reaper cannot meet. What if remote warfare undermines just war theory at a more fundamental level? As discussed, just war theory might conclude that a combatant has given up the right not to be killed. Perhaps a more thorough just war account would insist that a combatant has given up the right not to be killed only by certain people, namely, by opposing combatants. And what if the Predator or Reaper crew, because they remain several thousand miles from the target area, fail to meet the just war definition of a combatant? If so, then remote warfare fails to fit into just war principles, but not because the Predator and Reaper crews can make mistakes that ultimately target the wrong people. Instead, on this view, even when a Reaper crew targets only enemy combatants and only enemy combatants are harmed in the attack, the crew has nevertheless failed to meet the demands of just war theory.[13]

More generally, the question at the heart of this disagreement is the following: If I am morally permitted to kill an enemy combatant during war, is my moral permission based *solely* on facts about who the enemy combatant is and what he is doing? Or does my permission depend on some facts about me—for instance, where I am, how much risk I expose myself to, or whether I'm on the ground or in the air or on another continent? I believe that the moral permissibility of targeting the combatant has far more to do with the enemy combatant's actions and far less to do with facts about me. To see why, let's look at a historical example that has nothing to do with the Predator and Reaper.

Lieutenant George Welch was a US Army Air Corps pilot stationed in Hawaii in December 1941. He was one of only five American pilots to get an airplane into the air during the Japanese bombing raid on Pearl Harbor. After an all-night poker game at the officers' club, Welch and his roommate heard gunfire and raced to their airplanes at nearby Wheeler Field. Welch managed to launch his aircraft, fight, land, refuel and rearm, and launch a second sortie—all while under heavy fire from enemy aircraft. All told, he would shoot down four Japanese aircraft—all the while still wearing his tuxedo trousers from the night before.[14]

Did the Japanese pilots do anything morally wrong that day? The two views on just war theory offer different answers to this question. In the traditional view, there is no moral difference between Welch and his Japanese pilot adversaries. Both are combatants at war. Both wear the uniform and flag of their country. Both submit to the dictates of *jus in bello*. Each is morally justified in trying to kill the other because of their relationship as combatants on opposing sides of a war. In fact, according to this view, the Japanese pilots are morally justified in trying to kill Welch even though decision makers in Tokyo are morally responsible for an unjust war. A Japanese combatant who submits to the *jus in bello* principles hasn't done anything wrong, on the traditional view, even if Emperor Hirohito and General Hideki Tojo have.

What about the Reaper crew? Various theorists have argued that the Reaper crew members are not like Lieutenant Welch, because they are so far removed from the battlefield. The argument goes like this: In traditional just war theory, there are two important requirements for Welch's moral justification for trying to shoot down the Japanese pilots.[15] First, they have attacked him and his fellow soldiers and airmen. Second, he exposes himself to risk by battling against them.

Both of these elements, many have argued, are necessary for Welch to be morally permitted to shoot at the Japanese fighters. In the Reaper crew's case, we are missing one of these elements. Enemy combatants, say, al Qaeda or Taliban fighters, do attack US soldiers, but because the Reaper crew members do not expose themselves to al Qaeda's and the Taliban's attacks, they lack the same moral justification for shooting at enemy combatants that Welch has.

There are good reasons to challenge this idea. Are we really to suppose that combatants are permitted to use force only to defend themselves and never to defend fellow combatants? One important shortcoming of this argument is that this view leaves Dakota to fend for himself. Suppose the crew had been able to find the sniper while he was still shooting at Dakota and the rest of the platoon. According to this account of what it means to be a combatant, the crew would have had to sit there, helpless, as US soldiers were being shot at. Just war theory would have required them to say to themselves, "Look, I wish we could help. But we can't fire on the sniper, because we don't face personal risk and so we are not real combatants."

As recounted in Chapter 1, in the early spring of 2002, Technical Sergeant John Chapman, a combat controller, had been left atop Takur Ghar by his SEAL team and fought Taliban and al Qaeda fighters for an hour. An unarmed Predator watched his every move and saw him take his final breath. Are we really to suppose that if the Predator had been armed with a Hellfire missile, the crew would still not be justified in killing Taliban fighters to support Chapman? This conclusion cannot be right. The Kandahar-based crew would have been morally justified in defending Dakota in 2012 if they had been able to, just as the Predator crew would have been morally justified in defending Chapman in 2002 if they had been able to. We should reject the idea that combatants are justified in harming the enemy only if they face personal risk.

There is a more recent modified version of the reciprocal-risk argument: a combatant can be morally justified in harming the enemy while not facing personal physical risk as long as *the combatant's side* faces risk. In other words, on this modified view, a combatant's moral justification for targeting enemy combatants is still based on mutual risk, but it isn't mutual *personal* risk. It's mutual *structural* risk.[16] We are inching closer to a view that explains the various cases we've seen. The structural-risk argument helps to explain what happened to Jesus and Dakota. Jesus might not have faced the physical risk of enemy fire, but Dakota did. And so because both were combatants on the same side, and because their side faced structural risk, Jesus and his pilot were morally justified in harming enemy combatants to defend Dakota. Likewise, if the Predator armed with the Hellfire missile had been on station at Takur Ghar, the crew would have been morally justified in defending Chapman. Because Chapman faced individual risk, as fellow combatants on his side, their whole side faced structural risk, which would have justified the Predator crew's causing harm to the enemy. Finally, as Lieutenant Welch maneuvered his P-40 Warhawk in and out of Japanese fighter formations in December 1941, he was morally justified not strictly because he faced risk but because the Japanese navy was attacking Pearl Harbor. There was combat risk to Welch's side—the US military now stood in a relationship of reciprocal, structural risk with Japanese forces.

So far, this view seems to explain the relevant cases, but there is a downside. If the morality of killing in war depends on structural, reciprocal risk, then the United States would not be morally justified in using remote warfare to defend innocent people if they are not Americans. For instance, during the 1990s, after years of civil strife and multiple realignments of political communities, Serbian president Slobodan Milošević was using military force in Kosovo in an attempt to quell the separatist movement there. Sometimes called "the butcher of the Balkans," Milošević had already directed Serbian forces to kill fifteen hundred civilians, and the Serbian Army had already displaced a quarter of a million refugees.[17] In October 1998, the UN Security Council passed a resolution endorsing a ceasefire agreement between the two sides. The Serbian military violated the agreement, however, killing another forty-five civilians in January.

Ultimately, in March 1999, NATO member states unanimously agreed to authorize Operation Allied Force—a bombing campaign designed to compel Milošević and his Serbian Army to cease its hostilities against the Kosovars. This case, and many other cases of third-party intervention, seem to be justified.

In these kinds of intervention cases, though, there is no reciprocal, structural risk—at least not at first. If the moral justification for killing in war depends on reciprocal risk—even if it is risk to one's own fellow combatants or fellow citizens—then the use of force to defend foreign soldiers or foreign civilians can never be justified. If we believe our moral justification for combat operations is based solely on whether our side faces risk, then we would have no moral justification for intervening to stop Milošević from committing crimes against humanity.

Or we might ask similar questions of the Rwandan genocide in 1994. Many believe that a significant military power, perhaps the United States, ought to have intervened to stop the genocide. But if we adopt a view according to which we are justified in resorting to force only when our side is at risk, then even as Hutus are slaughtering Tutsis at a rate of roughly eight thousand murders per day, US soldiers had no moral justification for killing the Tutsi assailants—not even to stop the genocide.

This analysis might sound like hairsplitting, like asking fanciful questions about angels on pinheads or victims on trolley tracks, but these questions lie at the heart of not only remote warfare ethics but also the ethics of war more broadly. What constitutes a combatant's moral justification for killing enemy combatants? The arguments I've presented so far all point to risk. It is the risk the combatant faces, or that the combatant's side faces, that generates the moral responsibility for harming enemy combatants.

There is a better explanation: focus on threat, not risk. Your thesaurus may call these two terms synonyms, but there is an important difference. When we think about the morality of war in terms of risk, we tend to focus on the person or group that is facing that risk.

Dakota and his team faced risk. The Kosovar civilians faced risk. Welch risked being shot down over Pearl Harbor. When we think about the morality of war in terms of threat, we tend to focus on the person doing the threatening. The sniper threatened Dakota, Milošević's army threatened the Kosovars, Japanese airmen and Lieutenant Welch threatened each other.

Imagine that you and I are talking in a coffee shop. An armed person bursts through the door, intent on killing you. You are morally justified in harming the person in self-defense. Indeed, if killing the assailant is the only way to defeat the threat, you are morally justified in the killing.

If we focus solely on risk, though, we focus solely on you. Because you face risk, you are justified in causing harm. But what about me? Am I morally justified in harming the attacker? I should certainly think so—I am justified in causing harm to defend you even though I am not at risk. Most philosophers writing about defensive harm agree that the reason you and I are both morally justified in harming the attacker is not that you face risk but because the attacker poses an unjust threat. The harm you cause is in self-defense, and the harm I cause is in other-defense. As long as they are discriminate, proportionate, and necessary, both actions are justified because both actions would defeat the unjust threat.

The morality of war also depends on threat. Again, suppose the crew had been able to target the sniper who was shooting at Dakota and his platoon. The sniper has given up his right not to be killed because he threatens Dakota. And so Dakota's moral justification for shooting the sniper is the same as Jesus's moral justification—both men are justified in killing the sniper to defeat the threat the sniper poses to Dakota and his platoon.[18]

In reality remote warfare usually involves unbalanced risk between the Reaper crews and those they target. And if we assume that the most important element of moral justification is risk, then we will likely end up with a confused and complicated account of the morality of remote warfare. If, on the other hand, we believe that the most important element is defense against an unjust threat, it is much easier to see why some uses of military power are morally justified and others are not. Whether the killings of Mehsud, al-Aulaqi, and Soleimani were justified depends not on the risk facing the remote warfare crew but on the threat each man posed.

In fact, we can take an even broader look at the ethics of war to see just how central this notion of other-defense is. Individual military actions can be justified not merely in defense of fellow combatants or local civilians. When military force is justified, it is almost always justified in the defense of the whole political community—either one's own or someone else's.

## IN DEFENSE OF THE POLITICAL COMMUNITY

It is a sobering thing to stand in Coventry Cathedral. The roof, the windows, and much of the interior structure are no more. In its own way, it looks like other ruins across England—like St Augustine's Abby, reduced to ruins in the sixteenth century, or Minster Lovell, dismantled in the eighteenth. But Coventry Cathedral did not fall into disrepair five hundred years ago, and it was not scrapped for supplies three hundred years ago. It was bombed in the German Blitz on November 14, 1940. It is not an ancient ruin but a modern one.

The RAF began fighting in the Second World War's European theater long before the US Army Air Forces did. In the summer of 1940, the German Luftwaffe began a sustained campaign of aerial attacks against targets in the United Kingdom. The bombed-out cathedral at Coventry is just one testament among many of the destructive force of the raids. If, when we consider the RAF crews who fended off the Luftwaffe's raids, we think only of the aircraft in the air, then we will likely think only about self-defense justifications for killing in war. Or perhaps we might think of RAF crew members defending one another. Each RAF pilot attempted to shoot down German pilots to save himself and to save his fellow airmen from being shot down. But the moral justification for the RAF airmen's actions is much broader.

They didn't fight just for each other; they fought for the people of the United Kingdom.

With characteristic rhetorical force, Churchill reminds us in his famous 1940 speech that the RAF pilots accepted risks not to defend themselves or to defend the RAF but to defend the wider political community. "Never in the field of human conflict was so much owed by so many to so few."[19] We often talk about one state committing aggressive actions, or defending itself, or going to war. This is right, but incomplete. When a political community goes to war to defend itself, it will ask some subset of its population to shoulder that defensive burden. Some part will have to do the fighting on behalf of the whole.

Though the RAF crews faced the general threat the German bombers posed—alongside their noncombatant compatriots—by fighting the air war, they redistributed the threat, taking on considerable additional risk posed by the Luftwaffe fighters who tried to shoot them down. Churchill recognized that the RAF aircrew put themselves under greater threat so that their fellow citizens might be subjected to a lesser threat. "The gratitude of every home in our Island, in our Empire, and indeed throughout the world . . . goes out to the British airmen," Churchill said. When the RAF crews shot down Luftwaffe pilots and crews, they didn't do it primarily to defend themselves—though they did that too. Their primary purpose was to defend the British people.

We can think about most defensive wars in this way. When Japan attacked the United States at Pearl Harbor, Welch and others tried to shoot down Japanese airplanes and, later, to sink Japanese ships. Besides defending themselves, they were primarily defending the people of the United States against the threat of further Japanese attacks. And again, after the September 11 attacks on New York and Washington, US service members did not leave their homes and go to Afghanistan to defend themselves; they did so to defend their fellow Americans from further terrorist attacks by al Qaeda. People will disagree in particular cases whether this or that specific war was defensive and morally justified. My purpose here is not to referee those disputes. But if a war is indeed defensive, then combatants who participate in that war are morally justified in causing harm to forces on the unjust side to defend the innocent against the threat.

The same is true of remote warfare. If the war is just, then remote warfighters are morally justified in defending innocent people against unjust threats. And focusing on the threat brings more clarity to the moral questions than focusing on risk does. By focusing on the threat, we can more easily recognize that the remote warfare crew need not defend themselves and might not even defend friendly forces on the ground. In some cases, they might employ military force to defend US citizens at home. High-value targeting operations are morally justified only if killing the high-value target is discriminate, proportionate, and necessary. But this third *jus in bello* condition, the necessity condition, is met only if it is truly necessary to kill the target to defeat an unjust threat. I am not saying that all cases of high-value targeting are morally justified. Instead, I am saying that whether such an operation is morally justified depends in large part on whether the target poses an unjust threat. If so, then killing the target to defend the innocent people who would otherwise be attacked is the right thing to do, whether it's done with an infantry soldier's rifle, a traditionally piloted bomber, or an MQ-9 Reaper whose crew is half a world away.

This moral justification—the fact that killing a person on the other side of the world to defend the innocent can be the right thing to do—might help explain the positive emotional response many crew members had to their first weapon employment.[20] If we were to imagine only the Reaper crew and the terrorist leader on the other end of the weapon engagement as if they were two participants in a lopsided duel, a positive emotional reaction on the part of the Reaper crew would amount to a moral indictment. But if the Reaper crew is acting in the defense of fellow service members on the ground and fellow citizens at home—instead of on their own behalf, in their own defense—these positive responses are much easier to understand. If a

Reaper pilot is morally justified in targeting and killing a terror cell leader who is attempting to kill innocent civilians, the pilot not only does well, she does good.

Is everyone who poses a threat a viable military target? No, there are certainly other important constraints. Anwar al-Aulaqi, for example, was a US citizen. Surely, the United States owes to him as a citizen something that it does not owe to every operational terrorist leader around the world. Qasem Soleimani was no saint, but he was also a government official in a state with which the United States is not at war. Certainly, there are additional constraints on the use of force in such cases.

These are complicated questions, and reasonable people can disagree. For many people, the al-Aulaqi case came as a shock. For the first time since the Civil War, the US government targeted and killed one of its own citizens. Many people have probably also associated al-Aulaqi with his online religious persona rather than with his operational leadership role in al Qaeda. Of course I do not think the US government has a right to target US citizens (or anyone else, for that matter) solely because of online sermons, no matter how incendiary. But al-Aulaqi was more than an online preacher. Whether the United States was justified in killing him depends on whether a person can lose the rights of citizenship by waging war against his country. I think that an American can indeed give up his or her rights of citizenship. Title 8, Section 1481, of the US Code states that "a person who is a national of the United States whether by birth or naturalization shall lose his nationality by . . . entering, or serving in, the armed forces of a foreign state if such armed forces are engaged in hostilities against the United States."[21] By planning the Christmas Day airline bombing, al-Aulaqi had taken up arms against the United States. Although he did not do so on behalf of a foreign state, he did it on behalf of a nonstate actor. Even so, al Qaeda was a nonstate organization that had declared war on the United States.[22] If a person can lose citizenship rights for taking up arms against the United States on behalf of an adversary state, it seems plausible that a person might also lose these rights for taking up arms against the United States on behalf of an adversary terrorist organization.

Soleimani is a much harder case, which I will discuss in greater depth in Chapter 7. I suggest only that al-Aulaqi's American citizenship played an important role, as did Soleimani's position in the Iranian government. The United States didn't invade Iran or violate its airspace. But there is a sense in which killing Soleimani amounted to a violation of Iranian sovereignty—of using coercive lethal force to interfere in the Iranian state. Was that lethal force justified? It depends on the significance of the threat that Soleimani posed. Without access to the relevant intelligence, neither you nor I can answer that question with any confidence. But it is—or at the very least, it should be—much harder to justify killing government officials of other states with which we are not at war than it is to justify killing members of terror organizations with which we are at war.

The question that lies at the heart of the morality of remote warfare is whether a person can have a moral justification for killing a combatant from thousands of miles away. If we approach war as a contest that is largely defined by mutual, reciprocal risk for both sets of combatants, then the case for the justified use of remote weapons looks dubious. But approaching war in this way overcomplicates the question. We would do better to ask, first, whether the target poses an unjust threat and, second, whether killing them is both proportionate to the threat and necessary to defeat it. When viewed this way, the moral justification for battlefield violence has little to do with who faces risk and much more to do with who poses an unjust threat. If the US war in Afghanistan was a just war, then Tech Sergeant Jesus is morally justified in targeting enemy combatants to defend Dakota and his teammates—whether he flies overhead in a traditional aircraft, sits in a ground-based cockpit at Kandahar, or flies the airplane from the other side of the world. Predator and Reaper crews are also morally justified in targeting Taliban and al Qaeda leaders, as long as doing so is discriminate, proportionate, and necessary to defeat an unjust threat. The point here isn't that remote warfare carries with it some special moral justification—quite the opposite. The Reaper crew's moral justification for causing harm in war is precisely the same as the soldier's, the traditional pilot's, and the special operator's moral justifications—they are justified in causing harm to defeat an unjust threat.

# 5

# HUMAN JUDGMENT AND REMOTE WARFARE

W HEN WE THINK OF TRADITIONAL WARFARE, WE OFTEN SEE soldiers coming up with a creative solution to a tactical problem. On D-Day in Normandy, the Fourth Infantry Division landed at Utah Beach, more than a mile from their intended landing point. From there, success demanded improvisation. Brigadier General Theodore Roosevelt—son of the former president—famously announced to his soldiers, "We start the war from here." Well-known military aphorisms make the point more broadly: "The enemy gets a vote" and "No plan survives first contact with the enemy." In response to battlefield dynamics, soldiers have to improvise, and improvisation—coming up with creative solutions to problems—requires human judgment. The use of human judgment in dynamic combat environments can often have significant ethical implications. Even when soldiers are morally justified in taking life and causing harm, they have moral obligations to take as few lives as possible—to cause as little harm as they can to achieve the objective. Though some decisions to reduce the tragic moral costs of war will take place at headquarters staff and operations centers, soldiers in the field will often rely on their own judgment as they balance military objectives and moral costs.

Remote warfare is often described as push-button warfare. It's easy to imagine that the crew is relatively uninvolved in the decisions that influence the battlespace. We could imagine the president ordering a strike and a crew carrying it out without any need for discernment, ingenuity, creativity, or judgment. Indeed, some strikes might be like this. But even if that does happen on occasion, prescribed strikes of this kind do not capture the full spectrum of what remote warfare crews are asked to contribute to military operations. Crews are regularly asked to employ human judgment to achieve effects in the battlespace, to support and defend friendly forces, and to reduce as far as feasible the threats posed to noncombatants. If we just glance at the various organizational charts that govern Predator and Reaper operations, we might get the mistaken impression that the military hierarchy takes judgment out of the picture entirely. We might think of Predator and Reaper pilots and sensor operators as automata who simply translate directives from higher headquarters into actions.

In this chapter, I aim to show just how central human judgment is to remote warfare operations. Since crews respond to battlefield dynamics, they also respond to moral challenges on the battlefield. Imposing their own judgment is one way that remote warfare crews safeguard the morality of war. Because the Reaper, like the Predator before it, has the ability to loiter over the target area for hours on end, there is often time to observe activity on the ground and to build a case for striking, or for not striking, a target. As momentum builds toward a strike, the application of human judgment sometimes takes the form of a crew member pushing back against this momentum.

In March 2019, hours before the sun would rise over the Nevada desert, I sat in a ground-based cockpit behind a Reaper crew. Lieutenant William and his sensor operator, Staff Sergeant Sean, sat before an array of computer monitors showing maps, video feeds, system performance, and classified chat rooms. William talked to me in short bursts in between radio transmissions with an airman on the ground, half a world away. As described in Chapter 4, the joint terminal attack controller is the ground force commander's expert in all things airpower. Technically, JTAC is a joint certification, rather than an actual duty title. The JTAC is the person on the ground team who understands aircraft and weapons' capabilities and can legally say "cleared hot" on the radio, granting aircrew permission to release weapons in support of the ground force commander's intent.[1]

In between radio calls with the JTAC, Lieutenant William turned in his chair to talk to me. "One time, I saw a guy hunting in his front yard," William said, describing a Reaper sortie he flew over Afghanistan. "Everyone was spinning up, getting ready to shoot, because the guy had a weapon. But not everyone in Afghanistan with a weapon is bad." The rifle alone would not have been enough for the crew or the ground force to provide positive identification of the person under the crosshairs as an enemy fighter. The specific criteria depend on the rules of engagement for the Afghanistan area of operations, and rules of engagement are classified.[2] William was suggesting that a number of people involved in the mission saw the rifle as an indicator that the person was likely to be an insurgent fighter and not, say, a hunter. The momentum toward a strike was building. Analysts and ground forces watching the video would be watching for any other indications of insurgent behavior. William turned around in his chair again to look at me. "Eventually we saw him shoot a bird." Had it not been for William's confidence that the man with the rifle was not an enemy fighter, things might have turned out very differently, even tragically.

As momentum builds, sometimes releasing the weapon and killing the person under the crosshairs is the right thing to do. But as William saw with the bird hunter, sometimes the right thing to do is to push back against that momentum. William's ability to impose human judgment in the operation is crucial to understanding why the bird hunter was not killed that day.

In Chapter 3, we saw that even though the physical distance between remote warfighters and their weapons' effects might be great, their psychological distance can be much closer. But even this two-fold distinction is insufficient to capture the nuances of distance in war. Here I introduce a third concept of distance—the distance between weapons' effects and the application of human judgment. Ultimately, reducing the distance from which warfighters view the battlespace is about getting the important decisions right. Allowing crews, analysts, and commanders to see the war up close can enable them to use their judgment to decide how best to employ military force and how to minimize the moral costs of war.

Talking about the distance between judgment and weapons' effects might sound nonsensical. How are we supposed to locate a judgment in physical space? Imagine the globe of the earth in your mind, and map three weapons systems onto it—the B-2 stealth bomber, ICBMs, and the Reaper.

First, think of the B-2 Spirit bombers that launched from Whiteman Air Force Base, Missouri, and struck targets in Libya in the opening night of the 2011 NATO operation there, then returned to base.[3] In that thirty-six-hour mission, the bomber crews flew fifty-six hundred miles each way but delivered their munitions from relatively short range above the target. Second, think of the ICBMs that have been buried in silos in the American West for half a century. US Air Force missile officers, *missileers*, sit in fortified control centers awaiting an order from the president—an order that they hope will never come. If it does, they will begin the launch sequence on a weapon that will depart from Wyoming, North Dakota, or Montana and fly thousands of miles through air and space to strike a target on the other side of the world. Finally, think of the Reaper. The pilot and sensor operator remain in the continental United States, while the airplane flies over its target in Afghanistan. When the sensor operator aims the targeting laser and the pilot squeezes the trigger, the Hellfire missile will leave the aircraft and strike the target, again, on the other side of the globe.

Each system is long-distance warfare, but each is different from one another in important ways. The most obvious difference is between the B-2 and Reaper crews. Those who challenge remote warfare on ethical grounds argue that because the B-2 crew remains in the airplane relatively close to the target area, the B-2 stealth bomber action can be considered warfare. The Reaper crew, on the other hand, is often thousands of miles from the area of operations. But this physical separation—thousands of miles from the target area—is what the Reaper pilot and the missileers have in common. So what is the operational difference between an MQ-9 Reaper and a Minuteman III ICBM? Most important is the yield of the weapon. The Hellfire's thirty-pound warhead pales in comparison to the Minuteman III's several hundred kilotons.

But there is a second important difference. The Minuteman III crew remains in the United States, as does their perception of the operational environment. Their knowledge of the events that precipitated the launch—and the events that follow the launch—are limited to their nuclear-hardened control facility, buried underground in Wyoming, North Dakota, or Montana. Even if they were to learn something new while the missile is in flight, there is nothing they can do to recall, to render inert, or otherwise to interrupt the missile's parabolic flight path and, ultimately, its detonation. The security of the underground bunkers and silos, as well as the safety of placing these massive systems within the United States homeland, comes at the cost of flexibility.

In one sense, the Reaper crew is equally remote from the target area. But because of the airplane's camera and other sensors, and because the airplane is tied into the joint force network of air, space, sea, and ground systems, the crew members are not dependent solely on top-down orders. They can respond, in real time, to rapidly developing battlespace dynamics. Though their bodies remain in the United States, they have a perspective of the battlespace similar to that of aircrew in more traditional airplanes. As many have pointed out, thanks to the long sortie duration, their awareness of those battlefield dynamics is often better than that of their fellow aircrew in more traditional aircraft.

Focusing only on physical and psychological distance obscures questions about where human judgment can be applied by people operating in the battlespace remotely. If psychological distance refers to the effect the war might have on the crews, the concept of human judgment I have in mind here refers to the effect the crews might have on the war. Unlike many previous technological developments in which increases in physical distance meant that operators could not directly observe the battlefield, remotely piloted aircraft have vastly increased physical distances in warfare but tremendously reduced visual distances—operators can directly observe the battlefield.

What I am proposing here is a new idea I call the *judgment gap*, which, simply put, is the distance between the point of application of human judgment and the effects of that judgment. In Predator and Reaper operations, the judgment gap is smaller than it looks.

## JUDGMENT AND WAR

I left William and Sean and entered the cockpit of Lieutenant Briana, the pilot, and Airman First Class Calvin, the sensor operator, both of whom were flying their Reaper over Afghanistan. Briana had received a nine-line attack briefing from the JTAC. The *nine-line* is the standard attack briefing consisting of nine elements including, among other things, target coordinates, elevation, and a target description, that translate the ground commander's intent into specific guidance for the aircrew. In this case, the relationship between the aircraft crew and the JTAC was off to a rocky start. There are elements in the attack briefing the pilot was required to read back to the JTAC—the target elevation and coordinates, the location of the nearest friendly forces, and any specific restrictions for the attack.[4] Briana hadn't read back the briefing properly. Thanks to a garbled radio transmission, when the JTAC said friendlies were two kilometers southeast, Briana heard "two kilometers east." The JTAC told her that her read-back was wrong, but she had no way of knowing which item she had missed or how to correct it. When Briana repeated the read-back, the JTAC responded with emphasis to correct the error: "*South*-east." This might seem like a trivial miscommunication. From a tactical perspective, it was. Whether the friendly forces were two kilometers to the east or southeast, they were too far away from the target area for a strike to affect them. This conversation is nevertheless noteworthy because, even though they are separated by seven thousand miles, success or failure in execution can depend on the relationship between the crew and the ground forces. This relationship was off to a tenuous start.

The target the JTAC wanted Briana and Calvin to strike was a long, cylindrical object that showed up as a hot spot on the infrared video feed. The JTAC and his team had taken mortar fire from the area recently, and his motivation to destroy anything that looked like a mortar tube was understandable. But Briana wasn't convinced:

> I was working pretty heavily with the screener [an off-site intelligence analyst] to try to identify whatever the thing was. . . . That JTAC from my perspective was really [quick] to call any cylindrical hot spot a mortar tube. But you and I both know that a hot spot in [infrared], especially in a [high-definition sensor] ball, doesn't necessarily mean it's hot. It just means it looks different from everything else. The hot spot I found was out in a cultivated field inside a compound, and it was near [coordinate] grids where we thought the mortar tube would be.[5]

The JTAC's explanation for why the mortar tube would be in the middle of a field was that the mortar team must have buried it between uses to avoid detection. Briana and Calvin discussed this possibility and tried to determine if it had been buried. Through the high-definition infrared camera, an experienced crew can sometimes tell the difference between disturbed and undisturbed earth—a distinction that has been pivotal to Predator and Reaper crews since the early 2000s. As early as 2003, IEDs had become the greatest threat to coalition ground forces in Iraq and Afghanistan.[6] Crews were taught to look for disturbed earth along friendly transit routes as an indication of an IED buried in the road.

The JTAC can often see the video feed being produced and shared by the Reaper crew. In this case, however, he could not. His awareness of what the crew was seeing was based solely on what Briana told him. Briana conferred with the intelligence analyst, and ultimately, all agreed that the object was a shovel. Briana later explained their reasoning:

> It makes sense that a shovel would be lying out in a cultivated field. The farmer digs up some rows, puts the shovel down, and goes inside, right? And it was the middle of the night. . . . Why would they just lay a mortar tube out in the middle of this compound? Why wouldn't they hide it?[7]

The JTAC's insistence that the object was a mortar tube and, therefore, a viable target was not based on his ability to assess the object. Instead, it arose from his strong, and justifiable, interest in destroying anything that might be used to attack him and his team the following day.

In the end, Briana refused to drop a bomb on the shovel, and the JTAC put Briana and Calvin in what is often called the penalty box. That is, the JTAC directed them to hold in airspace out of the way and ignored Briana's radio calls for the remainder of the time I was in the cockpit. It is easy to think of this kind of response as petulant, but things looked different from the JTAC's perspective. He had had mortars lobbed at him for days. He was looking for support from a pilot and crew who had a responsibility to provide it. He asked the crew to drop a weapon on a target—a target he believed might be one of the very mortar tubes that had been shooting at him—and they refused. One can see why he might sideline this Reaper crew and search for another aircraft crew more willing to meet the intent of his ground force commander.

This episode was, in many respects, a nonevent. No weapons were released. No one was hurt. There were no explosions and no investigations. But it was significant. As Briana put it more than a year after it happened, it was "a hard sortie to forget." The account is significant for at least two reasons. First, the disagreement between Briana and the JTAC was a difference between two military professionals about what ought to be done. There was a judgment call to make, and these two disagreed about which call was the right one, given the circumstances. The sortie was especially memorable for Briana because, ordinarily, Reaper crews work hard to develop healthy working relationships with the JTACs they support. The conflict she experienced with the JTAC was uncomfortable and unforgettable.

Second, though the episode was a nonevent, it is significant in that it represents thousands of Reaper Predator sorties in which professional airmen act in good faith, applying all their technological and analytical tools, to do the right thing. In this instance, even some of Briana's peers in the Reaper community might disagree with her judgment. The JTAC, after all, was on the ground with intimate battlefield situational awareness. He and his team were the ones being threatened with mortar attacks—not an air force Reaper crew half a world away. Why should a Reaper pilot second-guess this JTAC's decisions? As a mere bystander, I was even further removed from the incident and was in no position to decide who was right. My intent here is only to show just how much human judgment is at work in the kinds of tasks Reaper crews are asked to carry out.

The question Briana, Calvin, and the JTAC sought to answer was not simply one about whether the long, cylindrical object was a shovel or a mortar tube. Instead, they sought to answer three interrelated question: What is it? How confident are we that it is what we think it is? And what are we going to do about it? Answering that last question is almost always a judgment call.

Understanding human judgment and distance in remotely piloted aircraft operations is difficult because the judgment gap doesn't always align with physical distance. Developments in military technology correlate with an increase in the judgment gap since there is often an increase in physical distance between warfighter and the effect of the weapons. There was, for instance, a marginal increase in the judgment gap in the oft-cited example of early remote weapons. King Henry V's longbowmen at Agincourt in 1415 were able to engage French knights at a distance. During the fleeting seconds that the arrow is in the air, the longbowmen do not maintain control over the weapon—they have no means of imposing judgment on where it will strike. Many military technological developments since Agincourt have followed this model: increases in physical distance result in increases in the judgment gap. The farther away you are, the less able you are to impose judgment in the battlespace—a trend that culminated with the ICBM.

Even so, any historical analogy is strained. Because the Predator and Reaper represent a new kind of warfare, there is, not surprisingly, no historical precedent that perfectly captures the nuances. These ideas about the judgment gap become clearer if we compare two US attempts on Osama bin Laden's life. In both, the physical distance between the warfighter and the target correlates with an increase in the judgment gap.

In 1998, President Clinton authorized a cruise missile strike against bin Laden after the al Qaeda bombings of US embassies in Kenya and Tanzania. The US Navy prosecuted the attack against what the United States believed to be bin Laden's location near Khost, Afghanistan, with ship-fired Tomahawk cruise missiles from the Arabian Sea.[8] Bin Laden had indeed planned on going to Khost, where he likely would have been killed in the strike. But, as Lawrence Wright describes in his Pulitzer Prize–winning book, *The Looming Tower*, bin Laden, in a car with his friends on the way to Khost, made a fortuitous decision:

> "Where do you think, my friends, we should go?" bin Laden asked. "Khost or Kabul?"
>
> His bodyguard and others voted for Kabul, where they could visit friends.
>
> "Then, with God's help, let us go to Kabul," bin Laden decreed—a decision that may have saved his life.[9]

In this case, a naval engagement control officer in the Arabian Sea was responsible for launching the cruise missile some five hundred miles from the target area. This naval officer, though considerably closer than the remotely piloted aircraft pilot thousands of miles away, still applied military force while remaining outside the theater of operations. But crucially in the bin Laden case, the engagement control officer had no means of imposing human judgment after the missile was launched.[10] This increase in the lag time between the application of human judgment and the weapon's effects—an increase in the judgment gap—can have a significant impact on whether the weapon achieves objectives and whether it generates troubling ethical implications.

In many circumstances, an increase in the physical distance between warfighter and target entails an increase in this lag time between the application of human judgment and the effect of the weapon. While King Henry's longbowmen accepted a marginal increase in the judgment gap, cruise missiles impose a much more significant increase. The missiles intended for bin Laden in 1998 could neither be recalled nor be redirected during the four- to six-hour flight time.[11] And, of course, even if they could have been recalled, the team on the ship in the Arabian Sea had no intelligence feedback loop to alert them to the fact that the intelligence reporting was mistaken. Though the physical distance was significant, the application of human judgment in response to real-time dynamics on the ground was completely absent. In this case, the judgment gap is correlated with physical distance.

Compare this 1998 event with the 2011 US raid that killed bin Laden in Abbottabad, Pakistan. President Obama opted for a capture-or-kill mission conducted by special operations forces that ultimately led to bin Laden's death. The forces in the helicopters and on the ground had both the capability and the authority to employ their judgment in response to real-time dynamics. The raid provides two important examples.

The first was the raid team's trust in human judgment. The first helicopter to arrive in the compound had planned to hover while the operators inside fast-roped into the compound. But the solid walls of the compound affected the airflow differently from how it did within the chain-link fence area where the team had practiced. The pilot had to put the helicopter down in the compound, ultimately in a forced landing that severely damaged the aircraft. The pilot of the second helicopter was unsure whether the landing and damage were caused by enemy fire or mechanical problems. In the face of this ambiguity, the second pilot decided to land outside the compound, forcing the SEALs to run in from there. Both the second landing and the direction of the special forces' approach were major deviations from the original plan.[12] In this instance, the team did not rely on scripted orders from higher headquarters or on calling back to higher headquarters for updated guidance. They employed human judgment.

Second, and more importantly, the room from which US cabinet and other officials, including President Obama, watched the raid lost communications with the raid force for some twenty to twenty-five minutes—over half the time the team was on the ground.[13] During this crucial period of the kill-or-capture mission, the raid team chose, in light of real-time dynamics on the ground, to kill rather than capture bin Laden. Again, they relied on human judgment. Leon Panetta, then director of the CIA, told reporters, "It was a firefight going up that compound. And by the time they got to the third floor and found bin Laden, I think it—this was all split-second action on the part of the SEALs."[14]

In each of these two cases, increases in physical distance correlate with increases in the judgment gap. Physical distance makes it harder to impose human judgment in the battlespace, and reducing physical distance makes it easier. The naval engagement control officer responsible for the 1998 cruise missiles was physically five hundred miles away from his intended target, and the point of application of human judgment was at his physical fingertips. His ability to react and to apply his own judgment in response to real-time dynamics was constrained by the technological limitations of the weapon and by the officer's great physical distance from the target area. The special operators in the Abbottabad raid, however, could perceive re-al-time battlefield dynamics and use their own judgment in response because, among other things, they were physically present in the target area. In these cases, increased physical distance entailed an increase in the judgment gap. But as we will see, Predator and Reaper operations tend not to widen the judgment gap.

## THE JUDGMENT GAP AND REMOTELY PILOTED AIRCRAFT

At first glance, the judgment gap in remote warfare may seem similar to that of the cruise missile. Unfortunately, our opinions have been influenced by widespread misconceptions. We are often told that re-motely piloted aircraft are robotic and that they fall into the class of autonomous or semiautonomous weapons.[15] *Robotic* and *semiautonomous* are better descriptors for the cruise missile. It flies a planned route toward a predesignated target, and the human operator cannot intervene postlaunch. Neither claim holds for the remotely piloted aircraft.

As discussed earlier, Briana and Calvin certainly imposed judgment. In fact, the pilot and the sensor operator of the Reaper pitted their judgment against that of the JTAC. This application of human judgment in Reaper operations was a theme in the stories recounted to me by air force Reaper crew members. One pilot, Captain Andy, told me about a time that the JTAC was confused and disoriented while taking enemy fire. The US unit was conducting a village clearing operation, and one part of the unit had been separated from the rest.

> [The friendlies] were getting shot at. Both sides were, I think, seventy-five meters apart. We got a nine-line [attack briefing] to shoot friendly forces. The sensor [operator] was like, "Holy crap. This is just not right." The hairs on the back of the neck stood up.

The sensor operator on that sortie told me that he slewed his sensor to the location the JTAC had directed him to, but what he found there didn't look right:

These guys' tactics, the way they're dressed, the weapons that they're carrying—they seem way too organized [compared with] the type of enemies we're seeing in these villages, and I don't think these are the bad guys. I think these are the friendlies. But . . . they had been sectored off. They had been cut off from the main group, and that's why [the JTAC] thought they were bad guys.[16]

Amid the fog of war, the JTAC had passed an attack briefing to the crew, but rather than provide coordinates for the enemy's position, he had mistakenly passed coordinates for a part of his team. The good judgment of the crew prevented a catastrophe.

The sensor operator on that sortie told me that he has pushed back against the JTAC on multiple occasions. Ultimately, in each case, the JTAC agreed to delay the strike. The sensor operator went on to say, "It's a two-way process between JTACs and aircrew. JTACs can tell us 'cleared hot' all day long and give us orders to strike, but of course, as aircrew, we don't have to, because the weapon is ultimately our responsibility."[17]

An instructor sensor operator, Technical Sergeant Megan, put it this way:

There [have] been several situations where I would say the conversation between the pilot in command or the crew and the JTAC . . . is—I don't want to say heated—but they feel like this is what needs to be done and the crew [says], "We're not comfortable with that" for whatever reason. . . . At the end of the day, this is [the pilot's] weapon. This is our aircraft. This is what we're comfortable with doing, and this is what we're not comfortable with doing. . . . I'd say most of our crews are very good at standing up for that.[18]

I asked another instructor sensor operator, Master Sergeant Sean, if he had ever experienced a moral dilemma in the seat. He said,

I've had several [instances] where we weren't comfortable with a certain strike just because we were worried about civilian casualties and things like that. So we pushed back to the JTAC and ended up waiting, and lo and behold, we were able to eliminate the target in clear terrain with no civilian casualties.[19]

The resounding claims from the US Reaper crew members with whom I spoke suggest that they do have the capability to apply human judgment.

None of these quotations is strong enough to suggest that the remotely piloted aircraft crews can employ the same amount of human judgment that the special operators in the bin Laden raid could. One of the most significant differences between cruise missile and remotely piloted aircraft missions is the point from which the pilot observes the battlefield. To see a target through a targeting pod at twenty thousand feet does provide the remote aircrew with greater awareness than was available in 1998, when the engagement control officer fired the cruise missiles from the Arabian Sea. But the position from which the remotely piloted aircraft crew observes the war is still very different from that of the soldier on the ground. Retired US Army General Stanley McChrystal, former commander of coalition forces in Afghanistan, put the concern this way: "Because if you see things in 2D, a photograph or a flat screen, you think you know what's going on, but you don't know what's going on, you only know what you see in two dimensions."[20] This is a genuine limitation on remotely piloted aircraft, but it is also a limitation on traditionally piloted aircraft.

If the judgment gap in Predator and Reaper operations is neither like previous generations of long-distance weapons nor like traditional warfighters on the ground, perhaps a closer point of comparison is traditionally piloted aircraft. Though the recent development of remote aircraft has had profound impacts on physical and psychological distance, perhaps the judgment gap in air operations is more continuous.

I spoke with Captain Shaun and Technical Sergeant Megan in the cockpit while they flew an operational mission over Afghanistan. Shaun has experience both as a Reaper pilot and as an MC-12

Liberty pilot—an unarmed, traditionally piloted, propeller-driven airplane used for intelligence, surveillance, and reconnaissance. One day, while he was flying the MC-12 in Afghanistan, the numerous intelligence analysts and ground personnel watching his video feed identified two people emplacing an IED in a culvert under a road. The various participants in the operation started preparing an attack briefing for another aircraft. Shaun and his crew were not convinced that what they saw was an IED emplacement and repeatedly intervened in the momentum that was building toward a strike. In Captain Shaun's words, "it didn't feel right." Megan compared this momentum of various team members preparing for a strike to a boulder that picks up speed as it tumbles down a mountain. "It's a lot easier to stop it before it starts than it is to stop it once it starts rolling down the hill."[21]

Shaun said that on this day in the MC-12, he and the crew were able to stall the kill chain multiple times. The *kill chain* is the US military's shorthand for the dynamic targeting process, consisting of six steps, "find, fix, track, target, engage, and assess."[22] Shaun explained what eventually happened:

> The two people we were watching ended up walking up to two full-grown adults. Once we saw the relative size, we knew the two people we had been watching were kids. They [had been] pulling sticks out of a culvert to get the water to flow. If we hadn't stalled the kill chain, who knows what would have happened?[23]

This is undoubtedly a case in which the crew applied human judgment in the battlespace. At fifteen thousand feet above the battlefield, Shawn could observe it and intervene from that position. Had he been a soldier on the ground, he would have observed from a very different position. The fact that the two people were children would have been more obvious, and his ability to apply human judgment would have been even greater.

When I asked Shaun about the differences between his ability to apply human judgment in the traditionally piloted MC-12 and in the remotely piloted Reaper, he said interrupting the kill chain is easier in the Reaper because he is now responsible not just for the camera being used to identify the target but also for the weapon. When a JTAC tries to push Shaun and his crew to release a weapon against their better judgment, he feels empowered to refuse. "I can say, 'I'm the A-code [the pilot in command]. It's my weapon. My sensor operator doesn't like it. We're not doing it.'"

Despite the many ways in which remotely piloted aircraft have changed the character of war, the level of aircrew responsibility bears a striking resemblance to aircrew responsibility in more traditional aircraft. And in practice, enacting that responsibility requires that aircrew must apply human judgment from a perspective that's relatively close to the battlefield.

## EMPOWERING JUDGMENT

The first limitation on the remotely piloted aircraft crews' application of human judgment is their perspective on the battlefield, watching through a set of cameras that are ten, fifteen, or twenty thousand feet in the air. The second is organizational constraints on their autonomy. Organizational limitations are not a question of technological capability but arise from organizational culture, doctrine, and training. The technological capabilities—the visualization of the battlespace via high-resolution cameras in multiple segments of light spectrum; the long loiter times over the target area; and the integrated network of operators, intelligence analysts, and commanders—are necessary but insufficient conditions for applying human judgment in the battlespace. Commanders also have to encourage it.

For the last few decades, many Western military organizations, including NATO, have moved toward a concept of mission command. According to this concept, commanders issue mission-type orders with an emphasis on the commander's intent, "thereby empowering agile and adaptive [subordinate] leaders with freedom to conduct operations."[24] This freedom, which is fundamental to mission command, consists of the liberty to employ human judgment in the battlespace. Subordinate commanders, including pilots in command, retain the authority required to apply human judgment even in complex and difficult circumstances. A recurring, though not universal, theme in my interviews with Reaper crews was that commanders at the squadron level and above would support pilots' decisions when those pilots employed their own good judgment—and especially restraint—in the battlespace.

Though I have focused on the pilot as the individual responsible for the aircraft and its crew, the sensor operator is also employing human judgment. As a subordinate crew member, that often means voicing concerns, observations, and sometimes even imperatives to the pilot.

The sortie in which Briana and Calvin disagreed with the JTAC on whether the object in question was a shovel or a mortar tube was not the first time Briana and Calvin had flown together. Months before, they had flown their Reaper above a convoy of US vehicles that was taking enemy fire in Afghanistan. Their role, often called *overwatch*, was to maintain awareness of the situation on the ground, to include friendly and enemy positions, and to report that information to the JTAC. A second aircraft, an A-10 Warthog (often called a "Hawg"), with its massive thirty-millimeter antitank cannon with depleted-uranium rounds, was also assigned to support the JTAC and his convoy. In the midst of the fighting, Calvin saw noncombatants trying to find safety from the gunfire. "I noticed that about ten women and kids were running away and into a building," he said. "I didn't think much of it then. . . . They probably ran when they heard the Hawg jet noise. They hadn't heard us, but they probably heard the Hawg."

Calvin didn't need to think much of it then, because, in searching for safety, the women and children had retreated to a building that was unrelated to the firefight. The noncombatants had found safety, and Calvin could focus on the friendly convoy and the enemy fighters.

As the situation developed, however, the JTAC—under the fog of war—became convinced that the fire he and his convoy were taking was coming from the very same building into which Calvin had seen the women and children run. Viewing the world from some twenty thousand feet above the battlefield, Calvin could have seen if there had been small arms fire from the building, but he had seen no fire from there. Calvin was sure of it. The JTAC passed a nine-line attack briefing to the Hawg, directing the pilot to employ his gun against what the JTAC still believed to be the enemy position in the building.

Briana worked with the intelligence analysts to try to get an updated account of who might be in the building. The intelligence personnel at the Distributed Common Ground Station can rewind the Reaper video to produce better analysis of a specific event. Surely they would have been able to see the ten women and children running into what was now the target building, but they couldn't give Briana an answer quickly enough.

Though she hadn't seen the women and children enter the building, Briana trusted Calvin. She keyed the mic and contacted the JTAC on the frequency the JTAC used to talk to all airborne aircraft, where she knew the Hawg pilot would also hear. She alerted the JTAC to the women and children Calvin had seen enter the building. The JTAC responded, "Copy; continue," meaning that he heard and understood what Briana had said, but that the Hawg pilot should continue the gun run.

It is very unlikely that the JTAC believed that there were women and children in the building and still chose to prosecute the attack. The much more likely explanation is that the JTAC believed his own account of the battlespace that he had perceived with his own senses. In the fog of war, he believed he was being attacked from that building, and his motivation to attack the people in the building to stop them from shooting is understandable. In other words, the JTAC probably trusted his own sense of the battlespace and rejected Briana's and Calvin's.

Though the JTAC wasn't dissuaded by Calvin's observation, the Hawg pilot was. He aborted the run and said he wasn't comfortable shooting at the building. Most likely, Calvin, a junior airman seven thousand miles away, saved the lives of those ten women and children that day. Had he doubted himself a little more or had valued his own judgment relative to the others involved a little less, he might not have spoken up and pressed the issue. Traditional warfare by itself—an A-10 and a JTAC—without the support of a remote warfare crew, would have produced a very different result. Calvin employed human judgment in the battlespace. As Briana would later put it, "That man undeniably saved some innocent people from harm and was the coolest head in the stack that day."

War—even remote war—remains a human endeavor, and remote warfare crews still employ human judgment. Though their physical distance from the target area is measured in thousands of miles, the distance that is relevant to human judgment—their judgment gap—is much smaller.

Practical wisdom or prudence—phronesis—is one of the virtues on which remote warfighters must rely. Others that must be cultivated, though they might seem counterintuitive for Predator and Reaper crews thousands of miles away, include traditional warrior virtues of courage, honor, and loyalty. And it is to those virtues that we now turn.

# 6

# IT'S "HARD WORK TO BE EXCELLENT"

## REMOTE WARRIOR VIRTUE

Is the Reaper a coward's weapon? Some certainly think so.[1] It isn't hard to see why. When we think of soldiers at the Somme or sailors at Midway, we can imagine the courage it must have taken to stand before the enemy and risk their own lives. But if Reaper crews do their fighting from the other side of the world, their actions seem anything but courageous. So, are those pilots and sensor operators cowards?

Our discussion of ethics so far has focused on actions: Is an action morally prohibited, permissible, or obligatory? But morality is concerned not just with actions but also with the character of the actor; not just with the things someone does, but also with the kind of person that someone is. For instance, when we raise our children, we want them not just to do the right things, which they might do from a sense of compliance or from fear of punishment. We also want them to become people of character—to become brave, generous, honest, and patient. Military organizations likewise want their members to have character virtues. For instance, the US Army's core values include loyalty, honor, and personal courage, and the US Navy and US Marine Corps share core values that also include honor and courage. The US Air Force wants its service members to put integrity first. The study of character virtues in moral philosophy is called *virtue ethics*, and it is as old as Western philosophy, tracing its history right back to Socrates, Plato, and Aristotle. And from that early start, virtues such as courage were considered crucial for warriors.[2] The character virtues particularly relevant to war are often called the *martial virtues*. One question we should ask about remote warfare is whether a remote warfighter can cultivate these martial virtues.

We can approach questions about martial virtue from one of two directions. In one sense, virtue is about morality. This is certainly what Aristotle had in mind. People who are virtuous live the good life—they act well because their actions flow out of their virtuous character. But when we talk about martial virtue, we often talk in more utilitarian terms, focusing on character virtues as a means by which we might achieve some good end. A soldier needs to cultivate the virtue of courage to complete a mission in spite of great risks. A sailor needs to cultivate the virtue of loyalty to serve shipmates well and even to put their needs ahead of the sailor's own. But if we talk in these utilitarian terms, we might inadvertently slip from a discussion of morality to a discussion of job requirements. We might fall victim to a bait-and-switch, using morally loaded terms like *virtue*, *character*, and *courage* but all the while talking only about vocational skills.

Joe Johnston's 2011 film, *Captain America: The First Avenger*, gives us a picture of both moral virtue and martial utility. Unassuming New Yorker Steve Rogers is a person of character. He doesn't like bullies. He's willing to stand up to defend the innocent even at great risk to himself, but he lacks the physical characteristics we often associate with soldiering. He's small and physically weak and struggles to meet the army's physical standards. Yet Rogers has an "it" factor: he has a virtuous character. The fictional 1940s US Army in the film can simply inject him with super soldier serum to overcome the physical challenges. Rogers is a hero in the classic comic book sense. He is a good man, and we want him to succeed. But the army leaders in the film see his character traits—his virtues—as the means by which he'll be able to accomplish his missions.

Despite the potential ambiguity between character virtues and vocational skills, questions about martial virtue or military virtue are central to the ethics of remote warfare. These questions are important because if remote warfare crews can take human life without exposing themselves to physical risk, then they don't need to cultivate the virtue of physical courage. But theorists have questioned whether remote warfare crews can cultivate the other martial virtues too, virtues such as loyalty, honor, and mercy. Fundamentally, questions about virtue ethics ask whether remote warfighters can carry out their duties and still be people of character.

Technical Sergeant Megan, an instructor sensor operator, told me about an overwatch mission she had flown for a US convoy in Afghanistan. The convoy planned to move through a village, but one of the vehicles had been damaged when the driver inadvertently drove off the road in the middle of the night. The convoy was stopped while a maintenance team attempted to fix the broken truck. Just before dawn, a group of people from the village had gathered to watch the event. They had no visible weapons, but, in Megan's words, something "just seemed off." The fact that a crowd gathered near the vehicles didn't bother her. Imagine a military force driving through your neighborhood and stopping outside your house. You would probably step outside to take a look. What seemed strange to her was that people further into the village—people who couldn't even see or hear the convoy yet—were looking in the direction of the convoy's approach. It was as if they knew the Americans were coming—as if the villagers were waiting.

Shortly before the vehicle was fixed, it was time for Megan and her pilot to swap out with the next shift's crew. They briefed the incoming crew on all the relevant aircraft and mission information—the airspace, any issues with engine performance, and radio frequencies and secure phone numbers for the supported unit. They also passed along their concerns. Megan briefed the new sensor operator, saying, "Nothing has happened, but a lot of people are gathering, looking in the direction that friendlies are supposed to be coming from, and it just feels off." The pilot similarly told the incoming pilot that the hairs on the back of his neck were standing up. In describing the whole event, Megan said, your experience and intuition get you to a point where you just think, "It doesn't feel right," or "It doesn't add up."

Shortly after the shift change, the maintenance team fixed the vehicle and the convoy started up again toward the base. As the vehicles passed through the village streets, a man on a rooftop slowly stood up and raised a rocket-propelled grenade to his shoulder. He fired.

The crew who had taken over for Megan and her pilot saw the shooter and fired a Hellfire missile at him before he could fire a second round. The first round had missed, and the convoy was unscathed. The Americans continued along the route and ultimately returned to base safely. But Megan couldn't have known any of this. She hadn't seen the man with the rocket-propelled grenade while he had been getting into position, and by the time he fired his weapon, she had already gone home. But she still had experienced that uneasy feeling that something wasn't right about the convoy. "I had anxiety for the next day or day and a half until I knew the whole convoy had gotten back safely," she told me. "You kind of feel a level of responsibility. . . . It's our job to keep them safe."[3]

This sense of responsibility she and her pilot feel for the friendly ground forces under their aircraft: Is it loyalty? Some have argued that it is impossible—or at least very implausible—for remote warfighters to cultivate virtues such as loyalty, courage, and honor. This issue, like others discussed earlier, is difficult to analyze without paying close attention to how remote warfare is new but at the same time similar to previous methods of warfare. On the surface, it might seem obvious that the US soldier assaulting Omaha Beach can culti-vate those warrior virtues in a way that the Reaper crew, thousands of miles from the target area, cannot. But there is more to the question of remote warfare and warrior virtue than distance.

The virtue of courage arises often in discussions about remote war-fare. The charge that these remote weapons represent a cowardly way to fight a war is all too common. It has emanated, as one might ex-pect, from enemies of the United States who have had to contend with the Predator and Reaper for many years. For example, Abu Mosa, a press officer in ISIS, was unambiguous. "Don't be cowards and attack with drones," he said. "Instead send your soldiers, the ones we humiliated in Iraq."[4] But accusations of cowardice among remote warfare crews are not limited to US adversaries. Some have identified a "narrative of American wickedness and cowardice," a stigma that remote warfare is "both illegal and a coward's weapon," and therefore assert that these aircrew members must also be cowards.[5] Some peo-ple in the US military, and even in the US Air Force, have had sim-ilar reactions to the growing Predator, and now Reaper, community. For example, a Pentagon army officer suggested that the government should not "honor people whose only hardship is spilling hot coffee in their lap as they move their joystick to fly a drone."[6] Retired air force colonel M. Shane Riza published a book in which he argued that on the most fundamental level, to kill from a distance is to kill "without heart."[7]

There have also been formal arguments about remote warfare and virtue.[8] According to one such argument, the martial virtues serve as a kind of standard against which warriors ought to be measured or an ideal to which warriors ought to aspire. From this point of view, a failure to meet the standard, or an inability to strive for the ideal, is a moral failure. Carrying this argument further, because the Reaper crew is not present in the battlespace with their fellow combatants, the crew lacks the context that is required to cultivate these martial virtues.[9]

One difficulty in these discussions is that it isn't obvious which character traits should count as martial virtues. These character traits must meet two conditions. First, they must enable the virtuous warfighter to overcome natural human tendencies that would inhibit mission accomplishment in war, and, second, they are character virtues in their own right. In other words, for a character trait to count as a martial virtue, it must be martial and it must be a virtue.

The first criterion is the virtue's applicability to the combatant at war. Consider courage, the archetype of martial virtues since Plato some twenty-three hundred years ago. If warriors are expected to fill their post even when faced with the very real and imminent possibility that it will cost them their lives, they must rely on a trait that enables them to act despite their fear. Courage enables virtuous warfighters to accomplish the mission in the face of their fears.

The second criterion is that a martial virtue must also be a moral virtue, or a character virtue, in its own right. Just because some trait or ability is relevant to combat does not mean we should include it in the list of martial virtues. Some soldiers can run faster, see better, or carry heavier things than others can. These traits might make them more effective on the battlefield, but running fast shouldn't be included as a martial virtue, because it is not a human virtue—it is a trait but not a character trait.

This is not a debate over mere semantics. There is an important moral question at stake here: Can remote warfare crews be good people? Many of the debates among military ethicists are about whether a military action is morally justifiable. Was killing a US citizen, Anwar al-Aulaqi, for example, morally justified? Or does the United States have a moral justification for killing government leaders such as Qasem Soleimani? Arguments about virtue are different because they refer not to a person's actions but to a person's character. Combatants who cultivate the martial virtues can kill without doing irreparable harm to their character.

The claim that remote warfare crews cannot cultivate the martial virtues has dire implications. Soldiers are asked to carry out violent

actions on behalf of the political community. These martial virtues—
courage, honor, and loyalty, for example—help soldiers maintain
their humanity in the face of war's brutality. And the opposite of virtue isn't just the absence of virtue. It's vice. If someone doesn't have
courage, she is either rash or a coward. If someone doesn't have honor,
he is dishonorable. These are ethical questions about whether Reaper
crews are necessarily of poor moral character. If remote warfighters of
the twenty-first century cannot cultivate these martial virtues, they
will lack the means by which combatants have always balanced the
violence they are asked to commit on the one hand against living
a full, flourishing human life on the other. The martial virtues are
crucial to understanding the morality of remote warfare because they
are central to whether remote warfighters can be people of character.

## VIRTUE ETHICS

The virtues are as old as Western moral philosophy. Though Socrates and Plato certainly discussed virtue, Aristotle offered the first
systematic account of virtue ethics in the fourth century BC.[10] And
remarkably, most virtue ethicists today adopt an approach similar to
the one Aristotle proposed more than two millennia ago. Contemporary virtue ethicists generally call themselves *neo-Aristotelian*. They
reject outright some of the premodern and illiberal views of the theory's founder. For example, Rosalind Hursthouse, who has been influential in the study of virtue ethics, suggests that neo-Aristotelians
"allow themselves to regard Aristotle as just plain wrong on slaves and
women."[11] But as a system that provides an account of what the virtues are, why they are important, and how they can be cultivated, the
neo-Aristotelians share a good deal in common with their namesake.

One reason that Aristotle's work on ethics took hold in Western
Europe was its influence on the Catholic Church, especially the work
of Thomas Aquinas. Aquinas studied at the University of Naples and
then the University of Paris in the thirteenth century—just as Aristotle was being translated into Latin for the first time in centuries.

A door to ancient Greek philosophy that had been closed to earlier generations was now open to the Dominican brother. In his 3,000-page *Summa Theologica*, Aquinas frequently refers by name to ancient and medieval philosophers and patristic fathers—Plato, Origen, and Augustine, for example. When he refers to Aristotle, however, Aquinas calls him simply "the philosopher."[12] Aristotle's influence on Christian ethics throughout Europe is grounded largely in Aquinas's frequent appeals to Aristotle in the *Summa*.

Virtue ethics, more than other available systems, is particularly relevant to people serving in the military. Because the Aristotelian system focuses on character rather than actions, it is well suited for ambiguous and unpredictable situations—situations that are common in the military. Given that warriors will face difficult moral dilemmas with life-and-death implications—and will perhaps face them with some frequency—they are in particular need of an ethical system that emphasizes practical wisdom rather than a set of rules or a mere determination of which action will produce the best outcomes.[13] The virtuous person not only can choose right rather than wrong actions in hypothetical cases but can also navigate the nuanced moral terrain of the real world. These virtues do, of course, apply to all aspects of life. Surely I should be loyal to my family members, honorable in my business practices, and morally courageous when moral obligations might come at a cost. These virtues are called martial only because those serving in the military can be expected to practice them with life-and-death consequences in the course of their duties.

Technical Sergeant Megan and her pilot can show us why virtue ethics are so relevant to remote warfare, but first, we will have to look more closely at virtue ethics. Megan felt responsible for the friendly forces on the ground and believed she owed them something. Was she motivated by the martial virtue of loyalty? To answer this question, we will have to look more closely at how the virtues are cultivated.

According to the Aristotelian and neo-Aristotelian systems, cultivating virtue has two important requirements. First, Aristotle's view had a clear end in mind: human flourishing. For Aristotle, flourishing is not a feeling but an "activity of the soul." A being flourishes when it cultivates the excellences—the virtues—that are appropriate for the kind of being it is.[14] Morality, to Aristotle, isn't just about doing the right kinds of things. It is also about being the right kind of person. And if we are living a flourishing life as a person, we will naturally do the right kinds of things.

Second, a person can cultivate these virtues by practicing virtuous behavior. Each time I am faced with the opportunity to act, for example, courageously, cowardly, or rashly, I have an opportunity to take a step toward cultivating either virtue or vice. We become "brave by doing brave actions," Aristotle says.[15] In the beginning, it will be difficult to act well. We act courageously not from a courageous state of character but instead because we know that we should act courageously.[16] It is unnatural and possibly even painful for me to act courageously at first. Over time, though, after repeatedly acting courageously against my inclinations, I can become trained to submit to the virtue. The person who has cultivated the virtue of courage no longer acts courageously just because it is the right way to act. This person acts courageously because it's who he or she is—a courageous person. This process, in Aristotle's view, is not an easy one.

The argument that Reaper crews can't cultivate the martial virtues is grounded in a simple assertion that they are not given opportunities to practice them. Think about the soldiers in Megan's convoy example. They live, eat, sleep, and fight side by side with their fellow soldiers. In so doing, they can practice acting out of loyalty, and ultimately, they can cultivate the virtue of loyalty, but does Megan have an opportunity to practice cultivating this virtue? She certainly does. She might be separated by seven thousand miles, but she can nevertheless look out for the soldiers' best interests, care about them, and do the very best work she can to protect them. Just because these virtues are of particular importance on the battlefield doesn't mean the battlefield is the only place we can practice them.

## MARTIAL VIRTUE AND MORAL VIRTUE

When we think about virtues such as courage, we might think of historical military examples: the Spartans at Thermopylae or Audie Murphy in France. But examples of heroic courage are by no means constrained to the battlefield. As crises and emergencies arise throughout history, there are always those who are willing to do what is right despite the consequences they might face: William Wilberforce, Harriet Tubman, Oskar Schindler, Martin Luther King Jr., Malala Yousafzai. Each rose to the historical occasion despite the very real threat of death for doing so. But if courage is relevant not just to combatants but also to activists, philanthropists, and leaders of political movements, then what can we say about the relationship between courage and remote warfare?

Scholars disagree about this relationship. According to one view, courage is a virtue anyone can cultivate. We saw this interpretation on display in London in 2018. Actor Benedict Cumberbatch, perhaps best known for his roles in the *Avengers* films and *The Imitation Game*, was in an Uber with his wife when he saw four men assaulting a bicycle delivery rider. Cumberbatch leaped from the back of the vehicle and dragged the assailants off the victim.[17] "They tried to hit him, but he defended himself," the Uber driver told reporters. "He seemed to know exactly what he was doing. He was very brave."[18] In the end, the attackers fled and Cumberbatch and the bicyclist embraced. This was neither war nor holocaust nor civil rights movement. It was an otherwise ordinary day on which Cumberbatch showed extraordinary courage. According to this first scholarly view of courage, anyone can cultivate it, and that's what Cumberbatch did.

There is another view, though. Some scholars distinguish between the courage shown by Tubman, King, and even Cumberbatch and the kind of courage required of military practitioners. According to this argument, Wilberforce and Schindler had courage, but they didn't have *martial* courage.[19] If this interpretation is correct, then the problem for remote warfare is that, unlike other combatants, remote crews cannot cultivate this important martial virtue, and so perhaps they are different in some morally significant way from traditional combatants. But this conclusion rings false.

In the historical cases, and even in the Cumberbatch case, people showed courage even though the consequences of doing so might be physical harm or even death. Of course, remote warfare crews don't ordinarily face physical risks like this. But even so, they might have opportunities to practice moral courage—to do what is right despite the consequences, even when those consequences are nonphysical. A number of Predator and Reaper stories throughout this book take this kind of shape: the ground team asks for a weapon, or leadership gives an instruction, and the crew members must decide whether to do what they believe is right in spite of the professional or social consequences. Now, you might think there is nothing particularly martial about that. Any of us might find ourselves in a situation in which doing the right thing comes at some professional or social cost, and when we do, we have the opportunity to show moral courage. And that's true. But I'm not sure this observation amounts to an argument against remote warfare. If the argument is that Reaper crews are not required to practice the same kind of courage—physical courage—as that needed among infantry marines, then we could all agree. But that contention by itself doesn't mean that Reaper crews are cowards. It just means that their opportunities to cultivate the virtue of courage will look similar to the opportunities that the rest of us face. This view, however mundane it may seem, fits better with the neo-Aristotelian account of virtue ethics.

We can practice the virtues—cultivating virtue through habituation—even in the small things. We can practice courage even when we aren't being shot at, don't see four aggressors trying to steal a bicycle, or aren't leading a political movement. We can practice loyalty even when there is no sniper involved. We can practice honor when what is at stake falls far short of life and death, and we can practice mercy every day with our children.

Courage has been considered a martial virtue for more than two millennia because combatants from Thermopylae to Kandahar have been asked to do a very difficult thing: they have been asked to expose themselves to considerable risk to achieve mission objectives. Courage is a martial virtue because, in practice, it is demanded of warriors. The virtue is a means, and the good is an end.[20] But it would be a mistake to think that this is the only kind of courage that counts as a character virtue.

Even if we all agree that remote warfighters can cultivate courage, loyalty, and honor, we should ask another question: Which virtues are especially relevant to remote warfighters? We should expect the set of relevant virtues to vary with different professional roles. It is clear enough that physical courage will seldom be demanded of Reaper crews, thousands of miles from their weapons' effects. But this does not mean that they will not have to show courage at all. Instead, we should expect there to be a set of character virtues that enable the remote warfighter to overcome natural human tendencies that would inhibit mission accomplishment in war. Remote warfare has redefined the warrior virtues—and this should come as no surprise.[21]

## THE MARTIAL VIRTUES IN REMOTE WARFARE

According to the neo-Aristotelian system, we develop a bad character if we repeatedly commit bad actions—and that's morally bad. We should instead endeavor to develop a good character. If remote warfare crews were constantly put in situations that required them not to act virtuously but instead to act viciously—cowardly, disloyally, or dishonorably—they would become disloyal, dishonorable cowards. Cultivating a vicious character would constitute a moral failure. But the performance of their duties does not require them to behave viciously. In fact, they have opportunities to practice virtue—even if that virtue might look different from that of the traditional warrior.

Some years ago, a Reaper pilot named Major Jordan and his sensor operator watched a Taliban training camp in an isolated mountainous region of Afghanistan.[22] This crew and others who helped make up the twenty-four-hour target coverage had watched the training camp for weeks in preparation for a large-scale joint attack on the camp. The attack would eventually include aerial strikes from US Air Force, Navy, and Marine Corps. The Reaper crew's job was tracking and recording patterns of life in the camp. After following a group of four individuals for a few minutes, the sensor operator told Major Jordan that he believed one of the four was a child or an adolescent. Jordan evaluated the video feed and agreed. The intelligence analysts—operating in a different part of the world from both the camp and the Predator crew—disagreed and reported the person as an adult male. Jordan insisted that it was a child, and the intelligence analyst insisted that it was not. The pilot eventually pushed his concern over the disagreement up his chain of command.

Meanwhile, more than thirty aircraft were marshaling to strike the camp. The Reaper squadron mission commander, who was responsible for all the squadron's active missions, pulled the video files from the cockpit hard drive and sent them further up the chain of command. Not long after, the senior US commander responsible for the joint attack on the camp reviewed the video and agreed with the Reaper crew: it was a child. The commander "rolexed" the attack—delayed it by thirty-six hours—and the Reaper crews continued to watch the camp until another strike mission could be planned and executed. Jordan and his sensor operator continued to track the child. In the end, all the terror training camp buildings were destroyed except the building that housed the child; that building was spared.

Jordan and his sensor operator were never under enemy fire, and there was no reason for them to demonstrate physical courage. They did, however, push back against the momentum that had been building toward the strike. Surely this act required moral courage. Philosopher Jesse Kirkpatrick is one of few who has publicly argued that remote warfare crews can cultivate the martial virtues. "Moral courage," he writes, "is distinct from physical courage by the simple fact that it need not entail physical risk. For example, to do what is ethical may require risk to a soldier's reputation, financial security, career, psychological health, personal relationships, and so forth."[23] Every time a Reaper crew member does the right thing despite potential risks to reputation, career, or standing in the squadron, he or she has an opportunity to act according to moral courage.

Loyalty, too, is relevant to remote warfare. As in Megan's case, Predator and Reaper crews must remain loyal to the ground forces they support as well as to their fellow aircrew. According to one common narrative, combatants are motivated not only by self-preservation or the abstract good to be achieved by winning the war but also by their comrades. Perhaps as long as there has been war, there have been combatants who return home to say that "it's about the person next to you." J. Glenn Gray writes, "The fighter is often sustained solely by the determination not to let down his comrades . . . Such loyalty to the group is the essence of fighting morale."[24] Though this phenomenon is more obvious in the traditional warrior's case, it still applies to remote warfare. Air force officer and pilot Dave Blair coauthored an essay with Karen House, a counselor to remotely piloted aircrews. Blair and House offered a depiction of why loyalty matters from the crews' perspective:

> In close air support, defending friendly forces in contact with the enemy, you hear the emotion in your ground comrade's voice, and you often hear the firefight in the background. . . . When the crew is able to save the lives of comrades and partners on the ground, this is as close to un-alloyed good as possible. . . . We act on their behalf, and when we can act effectively to bring about a good outcome, it's a good day. This is our culture, and it's one that focuses on moral agency while acting on the behalf of others.[25]

The Reaper crews' loyalty to ground forces indeed seems instrumental for mission accomplishment. As we have already seen in Chapter 3, a number of Predator and Reaper crew members have expressed deep concern for fellow comrades on the ground—even as they remain thousands of miles away.

Remote warfighters also rely on the virtue of loyalty among their crew as well as outside the squadron as they have always done since the advent of crewed aircraft. Predator and Reaper crews have likewise inherited a rich history of multiship operations.[26] One aircraft can zoom in to strike a target with the best precision possible while the other remains zoomed out to stay vigilant for noncombatants entering the area. Sometimes, one aircraft can provide the laser terminal guidance for the other aircraft's weapon. Or one pilot can quarterback a complex strike involving multiple aircraft with different weapons and different desired points of impact. Just as the fighter pilot in air-to-air combat relies on wingmen, so too do Predator and Reaper crews depend on their fellow squadron members flying other aircraft in the same airspace. And as we saw in Megan's case, squadron members must count on members of the squadron from shift to shift as well. Megan had to entrust the safety of the ground forces for whom she felt responsible to the sensor operator who took her place. One important difference is that the Reaper crew doesn't need to work toward self-defense and the defense of wingmen as fighter pilots sometimes do. But Predator and Reaper crews do pursue mutual support and tactical excellence to bring combat airpower to bear on the ground, as strike and multirole fighter aircraft have done for a hundred years. Whether the situation is within a squadron, between aircraft, or between remote crews and the friendly forces they support, loyalty seems to be an abiding martial virtue for remote warfare crews, and it has been, ever since these aircraft have been put to use.

Honor is a social virtue. One cultivates honor when one acts according to the norms of the social group. An honorable marine knows what it is to be a good marine. The honor in question is partly defined by the marine's own actions but not by those actions alone. People behave honorably only when their actions conform to what is widely understood to be the norms of their social group, which in this discussion is their fellow marines.

But conforming to social norms is not quite enough to define honor as a character virtue. For example, when a member of the

Mafia murders someone in a rival crime organization, surely the murderer behaves according to the group's social norms. When Michael Corleone pitches the idea of killing both rival Sollozzo and corrupt police officer McCluskey in *The Godfather*, one member of the Corleone family, Tom Hagen, warns that killing a police officer violates a long-standing Mafia rule. According to the rules of Corleone's social group, killing a leader of a rival Mafia family is permissible—even honorable—but killing a police officer is not. Michael will undoubtedly receive praise from his family members when he kills Sollozzo, and he will likely feel a sense of honor. But this sense is different from the marine who behaves honorably. Unlike the marine, Michael Corleone is in a social group that values the wrong things.[27] The last necessary ingredient is the choice of a social group. If you have cultivated the virtue of honor, you know which social groups should define honorable behavior. You must choose your peers wisely.

Though Predator and Reaper crews are often far removed both from friendly forces on the ground and from the civilians who occupy the area of responsibility, there are nevertheless two important social groups that can help to ground crew members' sense of honor. The first is the flying squadron, and the second is their fellow citizens.

Whether the flying squadron will have a virtuous or vicious effect on a crew member's ability to cultivate honor will depend on its culture. Which behaviors will be praised by superiors and peers? Which behaviors will be condemned? Though the precise number has varied over time, in 2020, the air force had twelve active-duty combat Reaper squadrons and another thirteen in the US Air National Guard. Squadron cultures will undoubtedly vary. The air force and the other branches of service have to balance the demands of a hierarchical organization on the one hand against the value that each crew member brings to the operation on the other. The pilot in command ultimately has the authority over the aircraft and its crew, but empowering lower-ranking members of the crew to provide inputs is crucial to mission effectiveness. Developing a squadron culture in which any member is empowered to speak up and point out potential failures is also crucial to cultivating honor. Technical Sergeant Megan put it this way:

> At the end of the day, I'm a sensor operator. So my pilot in command always has that final authority. That being said, the crew dynamic is that I have an open voice. . . . I have no problem speaking up and saying [something]. I know that they'll be receptive to it and take it into consideration before that final decision is made, and it'll be a discussion.[28]

If honor as a social virtue depends on the social group's praising some actions and condemning others, then the various members of the group must be empowered to provide input and to shape culture. Megan must have the institutional freedom, and indeed the responsibility, to speak up when another crew member is doing something that isn't right.

Sometimes the battlespace dynamics leave no space for a lengthy discussion, as was the case with another sensor operator, Staff Sergeant Jackson. When he was a junior airman and brand-new Reaper sensor operator, he and his pilot were tasked with targeting a high-value individual in a moving vehicle. The plan was to employ two laser-guided weapons, each set to a different pulse-repetition frequency code. The seeker system in a laser-guided munition is designed to look not just for any laser but for a specific laser spot being emitted in many short bursts at a predetermined rate. The laser is essentially flashing hundreds or thousands of times per second. The seeker is tuned to look for a laser spot that's flashing at a specific frequency. This setup allows two munitions from the same aircraft to track two different laser spots. In this case, the plan was to release two Hellfire missiles on two different pulse-repetition frequency codes to mitigate the risk of something going wrong with one of the weapons. As the sensor operator on board the aircraft, Jackson was responsible for maneuvering one of those targeting lasers. The other was maneuvered by a second aircraft on the opposite side of the target.

As the vehicle proceeded along the road, Jackson used an auto-mated track to maintain the crosshairs on the target. A track uses im-age processing software in the cockpit to grab on to a set of pixels on the screen. So, rather than hand-flying the crosshairs, Jackson was instead monitoring the system as it attempted to keep a lock on the vehicle. All was well. The pilot counted down to release, and the track held. With both weapons in flight, Jackson watched as the vehicle unexpectedly approached a building that would mask Jackson's la-ser. A third aircraft was responsible for zooming out to see the big picture and to provide the other aircraft with early warning of issues like this—collateral damage concerns or structures that would mask the laser. Jackson didn't know why that third aircraft had failed to alert him to this building in advance. If he allowed the crosshairs to continue to move in the direction they had been going, the laser spot would fall not on the vehicle but on the building. With only seconds remaining in the time of flight, there was no guarantee that the vehi-cle would emerge on the other side of the building before impact. If Jackson did nothing, the missile would hit the building—a building no one had been watching and that, in all likelihood, contained non-combatants. He recounted what happened:

> I . . . broke my track and guided [the crosshairs] around along the edge
> of the road, and [our missile] ended up missing because the vehicle was
> behind the building. The other asset's weapon that they were guiding
> hit. But if I had just left my track on instead of taking action myself to
> try to move the track out of the way, it probably would have drifted and
> then lased the building. And that missile could have potentially hit a
> building.[29]

Jackson and his pilot remained in position with eyes fixed on the tar-get area to provide battle damage assessment. As they watched, five civilians walked out of the building.

Most crew members would probably agree that under normal cir-cumstances, breaking the track and moving the laser off the target shouldn't have been Jackson's decision to make. He and the pilot, as well as the crew of the other aircraft, had briefed a plan and the pilot of Jackson's airplane had command authority over the missile, over the laser, and over Jackson. The question is not one of rank but of crew position. As a sensor operator who falls under the authority of the pilot in command, Jackson's professional obligation is ordinarily to submit to the plan as briefed. But this was not an ordinary situation. Jackson is empowered to act to meet the commander's intent. That is why, even though he deviated from the briefed plan, Jackson did not face repercussions for his actions. Instead, he was praised for having done the right thing under difficult circumstances.

This is just one anecdote, but it illustrates much of what I have seen in my own experiences and heard during discussions with other crew members. Though, as a military organization, the Reaper community does insist on a rank structure and hierarchy, the squadrons have generally cultivated cultures in which members are praised for doing what is right even when actually doing so means putting pressure on the hierarchy. This culture, by itself, does not ensure that crew members will cultivate honor, but it does create an environment in which the social group can praise some behaviors and condemn others. And if the behaviors being collectively praised are morally praiseworthy, then squadron members will indeed have the opportunity to cultivate honor.

In addition to the flying squadron, there has been, in post-9/11 conflicts, a second social group whose norms are perhaps even more important for the cultivation of honor than enemy combatants are. The US populace—the citizenry that commissioned US combatants to conduct violence on its behalf—seems at least as important a standard of social norms as enemy combatants and battlespace civilians. The combatant's fellow citizens are now probably a far more important barometer for the warfighter's honor. In every case but the Civil War, when Americans went off to combat, they packed up and left not just their families and immediate communities but their country too. In recent decades, when traditional aircrew deploy, they are surrounded by their wingmen, and many will have opportunities to interact with ground forces, but they will be dislocated from their communities and fellow citizens. The Reaper crew members, on the other hand, can measure their actions against the values of their fellow citizens. One trope in the remote weapons literature is that a Reaper pilot conducts a strike in theater and then attends a kid's soccer game an hour later. In addition to regularly attending family dinners and children's activities, remote warfare crews must also interact with friends, family members, neighbors, and acquaintances day in and day out. Though classification and security requirements prohibit crew members from sharing the details of their work with friends and neighbors, surely these daily interactions provide the service members with the opportunity to evaluate their own behaviors against the norms of an important social group—the society in whose name they act. In one sense, remote crews might be better able to measure their actions against the values of the community, and thereby to cultivate honor, than can their counterparts who are deployed for four, six, or twelve months on end.

## RESPECT FOR HUMAN DIGNITY

Major Richard "Dick" Winters, whose military service in the Second World War was made famous in *Band of Brothers*, once said, "War brings out the worst and the best in people. Wars do not make men great, but they do bring out the greatness in good men."[30] It's an apt summary of the martial virtues. The "good men" to whom Winters refers have cultivated the character virtues. When tested in war, they will be capable of doing great things. But Winters also offers a caution: war can bring out the worst in people too. In wars throughout history, combatants on the ground, in the air, or half a world away, have been tempted to see the enemy as less than human. Cultivating a respect for human dignity is perhaps the only way to defend against this insidious threat. While respect for human dignity is not a prerequisite for accomplishing the immediate task—assaulting the beach, attacking the machine gun nest, or releasing the missile—it is a necessary condition for warfighters' ultimate task of fighting to achieve just aims and returning whole.

Dehumanizing the enemy in war is not the evil work of a sadistic few. It is the natural position to which human character will sink without positive intervention. In his Vietnam War memoir, Karl Marlantes describes his emotional response after targeting enemy forces in defense of a friendly reconnaissance team:

> I didn't kill people, sons, brothers, fathers. I killed "Crispy Critters." It could have been krouts, nips, huns, broche, gooks, infidels, towel heads, imperialist pigs, yankee pigs, male chauvinist pigs . . . the list is as varied as human experience. This dissociation of one's enemy from humanity is a kind of pseudo speciation. You make a false species out of the other human and therefore make it easier to kill him.[31]

The list of epithets Marlantes mentions here is meant to tie his own act of dehumanizing the enemy to so many other instances in so many other conflicts. Cultivating respect for human dignity in combat has always been difficult. Yet, throughout modern history, many combatants have resisted temptation and cultivated a healthy respect for the humanity of the enemy.

Hans von Luck commanded a German panzer battalion in North Africa during the Second World War.[32] Having been discovered by British reconnaissance aircraft, the battalion was subsequently attacked by Allied aircraft. Nearly every vehicle in the battalion had been damaged, including the battalion's entire complement of surface-to-air weapons. With no defenses at the battalion's disposal, one vehicle remained. Luck shouted orders to his intelligence officer who stood next to the remaining vehicle. The officer passed the instructions to a radio operator inside the vehicle, who broadcast those orders to the rest of the unit. A Canadian pilot—Luck is not certain, but he believes it was a Canadian pilot—proceeded inbound on a rocket attack run to strike the last remaining vehicle; the attack would undoubtedly have killed the radio operator and intelligence officer. Instead of releasing his weapons, though, the pilot waved his hand at the two soldiers, encouraging them to leave the area. He leveled off and rolled into a wide turn to reset the attack. In the meantime, Luck ordered the two men out of the vehicle. On his second pass, the Canadian pilot struck the vehicle, leaving the two soldiers unharmed.

The Canadian pilot was under no legal obligation to spare the two German soldiers. They were enemy combatants and were transmitting Luck's military orders, that is, they had not surrendered and were actively engaged in military activity. Nor was there any apparent moral prohibition against killing them. They were combatants participating in an unjust war and had given up their right not to be killed. In the most literal of terms, the Canadian pilot had them *dead to rights*—they had no right not to be attacked by him. But instead, he acted independently, as a moral agent, and out of respect for their humanity. He acted as a virtuous person and as a warrior. "The attitude of the pilot," Luck later wrote, "became for me *the* example of fairness in this merciless war. I shall never forget the pilot's face or the gesture of his hand."[33]

It would have been another matter if the Allied aircraft had not already disabled Luck's battalion. The antiaircraft guns had been destroyed, and Luck and his men had no means of targeting the Canadian pilot. The Canadian pilot's action was not one of physical courage. As far as the Germans' weapons' range was concerned, the Canadian pilot might as well have been several thousand miles away. It is impossible to understand the Canadian pilot's actions without reference to this reduction in risk. It was the absence of enemy fire that created the time and space for him to act out of respect for human dignity. Perhaps the same is true of remote warfare crews.

Respect for human dignity was not limited to the Allies in that war. In May 1940, German fighter pilot Hans-Ekkehard Bob flew his Me-109 over northern France. After a lengthy and tiring engagement with the French Hawk pilot, Hans forced the Hawk to land wheels-up in a field, securing the fourth of what would eventually be a staggering sixty total air-to-air victories during the war. Historian and author James Holland described what happened next:

> Without a second thought, and safe in the knowledge that they were still in German-occupied territory, Hans lowered his undercarriage and touched down beside the Frenchman. . . . Having tended to his wounds he then took the pilot's name—Sergent-chef Bés—and promised to write to his parents to let them know he was safe. . . . "What I did was forbidden," says Hans. "I could have been court-martialed for that."[34]

Saving the French pilot's life was technically forbidden, yet Hans acted out of respect for the dignity of his fellow human. In fact, true to the Aristotelian system, Hans did so reflexively. Like Cumberbatch, he acted as one who had cultivated the relevant virtue without a second thought. He acted virtuously for virtue's sake. But notice that his explanation of the event suggests that he felt empowered to act virtually in part because he was still safe in German-occupied territory. His act of mercy, which grew out of his respect for the dignity of the French pilot, was contingent on reduced risk to himself.[35]

Perhaps the same is true of remote warfare. Perhaps remote warfighters will work under the same natural tendency to dehumanize those against whom they fight. Yet, like the Canadian pilot and Hans, perhaps the fact that they operated at drastically reduced levels of risk means that they will be better able to cultivate a healthy respect for human dignity—dignity even of their enemies. Perhaps, though their war will consist not of discrete deployments and campaigns but of exposure to violence for years on end, they will be able to cultivate this important virtue and return whole.

Aristotle knows that he's asking a lot. It's "hard work to be excellent."[36] The virtues relevant to traditional soldiers are relevant to remote warfare crews too. But in addition to the traditional martial virtues of courage, honor, and loyalty, remote warfare crews must also cultivate respect for human dignity. The Predator pilot who takes human life or who kills enemy combatants who have given up their right not to be killed, without ever considering human dignity or the conflict between justice and mercy, risks giving up too much of himself or herself to war. The oath US service members take includes what is often called the *unlimited liability clause*. They offer to give up their lives in support and defense of the US Constitution. They offer their lives, but they should not offer their humanity. The martial virtues for the remote warriors—including respect for human dignity—are the distinctively human character virtues that push back against combatants' natural human tendencies, including the propensities for callousness and apathy. In the absence of physical risk to self, respect for human dignity is not only genuinely possible for the remote warfare crew but also an indispensable bulwark against the moral harms of war. In any event, both the traditional warriors and the Reaper crew would do well to remember Aristotle's admonition.

# 7

# WHAT COMES NEXT?

I T IS NOW TIME FOR US TO PEER INTO THE FUTURE.

We have looked at remote warfare and assessed how it compares with the long history of war. We have also focused on the twenty-first century, and especially the twenty years of war in Afghanistan, to understand better what remote warfare is and what it means for the changing character of war.

We are at a historical inflection point at which the proliferation of remote weapons, a broadening scope of targeted-killing operations, and developments in AI will collectively shape the future of war. What will that future look like? And central to our discussion, what ethical lessons from the first twenty years of remote warfare are relevant to future conflicts?

Though the United States, the United Kingdom, and Israel enjoyed tripartite hegemony in their use of combat-capable remotely piloted aircraft, that lead has eroded. Such states as Russia and China are developing and exporting remote warfare systems, many other states are buying them, and such nonstate actors as ISIS have adapted small, low-cost drones for combat uses. What does this proliferation of remotely controlled weapons systems mean for the ethics of remote warfare in the coming decades?

The US has broadened the scope of its use of remote warfare to strike leadership targets—even leadership targets who would have seemed off-limits just a few short years ago. This expanded scope was perhaps most noteworthy in the January 2020 killing of Qasem Soleimani. Though the list of Soleimani's military activities against US citizens and interests is lengthy, he was an official in a government with which the United States was not at war. The killing of Soleimani represented a novel and potentially destabilizing use of remote weapons. Does this action somehow set a precedent for future strikes against government officials?

Throughout this book, I have insisted that Predator and Reaper aircraft are not autonomous. That is, they have no more sophisticated autopilot capabilities than do traditionally piloted modern aircraft. But this could change. The age of AI is upon us, and the US military is investing in not just remotely piloted aircraft but truly unpiloted ones. What can we expect from a future in which aircraft are not just controlled from afar but also enabled by AI? What lessons about psychology, human judgment, and martial virtue might still be relevant in that case?

## THE PROLIFERATION OF REMOTE WARFARE

Not too long ago, a drone-strike headline would imply a US Predator or Reaper. But those days are gone. Theorists gave early warnings about "drone proliferation."[1] We in the United States ought to be careful how we use these systems, the warnings insisted. We might be the only power with the capability now, or we may be among the few, but technological military advantages don't last forever. Eventually, we were warned, rivals and adversaries will also have the ability to employ military force from vast distances. What precedent will we have set?

In September 2019, the effects of this proliferation were visible on the world stage. Though some reports disagree on the details, two Aramco oil production facilities in Saudi Arabia were attacked by what commentators referred to as drones.[2] Until recently, most observers would have thought only the United States, the United Kingdom, or perhaps Israel could have launched such a coordinated attack by multiple aircraft some fifteen hundred miles inside Saudi airspace. In this case, the aggressor who claimed responsibility was not a major power but was Yemen's Houthi rebels—though some US officials suggested that the Houthis conducted the attack with Iran's support.[3] Even so, the remotely controlled aircraft strike as a foreign policy tool was no longer constrained to the arsenals of the few states that had introduced it. Both Russia and China have developed their own armed remotely piloted aircraft. And though the United States and Israel were the only major exporters of remote combat aircraft, China has become a significant actor in that role as well, exporting to, among others, Egypt, Iran, Jordan, Pakistan, Saudi Arabia, and the United Arab Emirates.[4]

The questions theorists raised a decade ago will now be answered in practice. Has the United States established practices and precedents that other states will follow? And if so, will these practices support the international order and the rule of law? The US use of force outside "areas of active hostilities" has, in the years since 2001, been defended by referring back to the 2001 congressional Authorization for the Use of Military Force (AUMF).[5] Not being a legal scholar, I wade into this discussion with caution. Even so, there are two points that have been raised by legal scholars. The first is that the AUMF established a scope for the congressionally approved use of force in response to the 9/11 attacks on the United States. The authorization was for the use of military force only against "those nations, organizations, or persons [the president] determines planned, authorized, committed, or aided the terrorist attacks . . . or harbored such organizations or persons."[6] In the 2012 National Defense Authorization Act, Congress expanded the scope of this authority to include al Qaeda's "associated forces."[7] This act clearly provides authorization for war against al Qaeda in Afghanistan and, later, against al Qaeda in Iraq (though it didn't justify the 2003 invasion of Iraq).[8] But we would be hard-pressed to show that, for example, ISIS, an organization that has from its inception opposed al Qaeda, meets the AUMF's standard. There are questions, therefore, about whether the United States submits to its own domestic legal constraints on the use of force. Second, international legal scholars have argued that, even though Congress passed the AUMF, the US application of remote warfare outside international armed conflicts violates international law.

As the technology that enables remote warfare becomes cheaper to produce and easier to acquire, proliferation is inevitable. In the years following the 9/11 attacks on the United States, the world's only superpower employed remote warfare as a foreign policy tool in ways that have established norms and expectations for others. If US power relative to other states is waning, as many scholars suggest, then we should look closely at the behavior of regional powers' use of remote warfare in the coming years. Time will tell whether other states will avail themselves of this ability to strike individual targets outside areas of declared hostilities. Likewise, only time will tell whether those states will take seriously the just war constraints of discrimination, proportionality, and necessity.

## TARGETED KILLING OR ASSASSINATION?

The United States has established other practices that might serve as potential future precedents as well. As discussed, Iranian general Qasem Soleimani was killed in January 2020, reportedly by an MQ-9 Reaper crew.[9]

In conversation, I have often defended the United States against the charge that its use of remote warfare to target high-ranking members of enemy terrorist organizations amounts to assassination. There are technical reasons for my argument, but there is also a more fundamental reason. We should use weighty terms like *assassination* consistently, or they risk losing their meaning. For instance, was the raid on Osama bin Laden an assassination? Surely there are some who think it was. But most commentators avoided the use of that loaded term, probably for many reasons. One obvious reason is that *assassination* carries with it a negative connotation. Assassinations are bad, and most Americans thought the killing of Osama bin Laden was justified.[10] Calling it an assassination would be at odds with how most Americans viewed it. Calling it an assassination also seems at odds with the relevant facts. Bin Laden was the operational leader of a terror organization that had declared war on the United States and with which the United States had been in conflict for a decade. The term *assassination* is often reserved for contexts other than war.[11] And if killing bin Laden in a special operations force raid is not assassination, then it doesn't seem right to call the killings of his subordinates assassination either, even if they are killed by Reaper strikes.

This isn't a recent distinction. In 1943, a four-ship of American P-38 Lightning fighter aircraft took off from the Solomon Islands for a 400-mile flight, just above the wave tops to Bougainville Island off the east coast of Papua New Guinea. There, they intercepted an aircraft carrying Japanese admiral Isoroku Yamamoto—the mastermind behind the Japanese attack on Pearl Harbor two years before. There was intense fighting between the American pilots and the Japanese Zero pilots escorting the admiral's plane. Yamamoto was shot down and most likely died in the ensuing crash. One of the American pilots was also killed. The event was surely an attack on a high-value target, but was it an assassination? Most theorists don't think so. The United States and Japan were two nations at war, and Admiral Yamamoto, senior officer though he was, was a combatant in that war.

The January 2020 Soleimani strike was different. Soleimani was a general in the Iranian military, and in that sense, he had something in common with Yamamoto. But unlike the United States and Japan in 1943, the United States and Iran were not two nations at war in

2020—at least not in any conventional sense. Considerable reporting suggests that Soleimani's organization, the Quds Force, had been engaged in widespread and systematic efforts to kill American service members. Soleimani's Quds Force has been cultivating relationships and raising militias across the Middle East for decades. A number of those militias have, according to reporting, engaged US forces directly in Iraq. The Trump administration attributed six hundred US service member combat deaths to Quds Force militias.[12] But that doesn't mean that the United States and Iran have been at war and that, therefore, Iranian military members are combatants and lawful targets.

We can ask at least two questions about the strike on Soleimani. First, was it morally justified, and second, was it wise? As discussed in Chapter 4, there are three criteria for the permissible use of military force in an ongoing war. These fall under the category of *jus in bello*. Is the attack discriminate? Is it the least harmful means of bringing about the good end? Is it proportionate?

The strike was directed against a person who bears responsibility for the Quds Force attacks on US military members. But Soleimani was not the only person killed in the attack. In response to the rise of ISIS in Iraq, the militias—many of them Iranian-supported—were organized into a broader organization called the Popular Mobilization Forces. Abu Mahdi al-Muhandis, the deputy commander of those forces, was also killed in the attack on Soleimani, as were several others.[13] Suppose that Soleimani was the only intended target. If so, the attack was discriminate. Soleimani's Quds Force was acting as an enemy of US forces, and Soleimani was its leader.

Was the strike necessary? That is, was killing Soleimani the only way, or the least harmful way, of preventing the Quds Force from attacking US service members? Perhaps. The answer to this question depends on information that the public (myself included) cannot access. US officials claimed at the time that Soleimani posed an imminent threat—that there would have been imminent attacks on US bases or embassies had it not been for Soleimani's death. Suppose we grant all these claims. The Quds Force was actively engaged in fighting against the US military. Soleimani was a uniformed officer responsible for the Quds Force. He was the only intended target. And he and his force posed an imminent threat so great that intentionally killing him, and killing several others as a side effect, was justified to prevent the harm. Suppose all this is true; there nevertheless remains an important question.

Was killing Soleimani proportionate in the long run? Unlike Osama bin Laden, Anwar al-Aulaqi, and Baitullah Mehsud, Soleimani was a government official. Though the United States did designate the Quds Force a terrorist organization in 2019, that didn't change the fact that Soleimani was a government official. And unlike Admiral Yamomoto, Soleimani was an official in a government with which the United States was not at war. If an assassination is a targeted killing against a political leader of a political community with which one is not at war, then the strike against Soleimani was an assassination.

When we consider the just war principle of proportionality and weigh the good to be achieved against the moral costs, we should consider long-term costs as well as short-term ones. By killing an official of a government with which we were not at war, has the United States opened the door to other states engaging in this kind of action? Have we established a precedent that states can use to justify this kind of action even against our own government officials? There is no direct analogy to Soleimani's position in the Iranian government, but we might think of the secretary of defense, the director of the CIA, or the commanders of elite special operations units. Was it wise to usher global powers into a world in which killing government officials such as these—even while not openly at war—will be seen as an accepted practice among states? Time will tell. Perhaps these concerns will amount to nothing more than empty alarmism. I hope so. But in any event, the proportionality calculus must include both long-term and short-term costs. And even if Soleimani was planning imminent attacks on US persons, we should be concerned about the long-term proportionality calculus in the strike that killed him and about the normalizing effect that the strike may have on assassination by remotely piloted aircraft in the future.

## ARTIFICIAL INTELLIGENCE

In addition to the proliferation of remote warfare and the increased scope of US strikes, the third, and perhaps the most profound, change coming to remote warfare is in AI. The Reaper, like the Predator before it, is not autonomous. Every weapon engagement requires the human crew's direct input. I have also argued that these aircraft are a techno-social system consisting of both a machine and a human crew. But in the near future, the relationship between the human and machine will change. AI is coming to remote warfare.

AI is a broad category whose definition is often debated, but we can think of it as a machine's ability to do that which used to be the sole purview of the human.[14] Under this admittedly broad definition, AI can be as rudimentary as computer memory. As drafts of this book approached eighty thousand words, I had no hope of remembering all of them. But my computer has no trouble with that task. In fact, it can store all the words, recall them at my request, and even show me which words are misspelled. If memory was once the sole purview of the human mind, and if computers can now "remember" better than we can, then this is a kind of AI—even if it has been around for decades. But computer memory is not the reason that AI has been such a frequent topic of conversation in recent years; nor is it why I bring it up in this final chapter.

The explosion in AI capability since around 2010 has taken place largely in one specific subdiscipline in the broad field of AI: machine learning. Because computers can process information very quickly, these machines can view thousands, millions, even billions, of bits of data and can identify patterns in that data. Machine learning algorithms are models that recognize variables across a range of dimensions and identify patterns in those variables. Imagine that I had all the health records for six million people—the number of veterans served by the US Department of Veterans Affairs.[15] If I were to pull out age and height from all those records, I could plot them on a two-dimensional graph. You and I would notice a pattern. For infants and children, height is correlated with age. In the aggregate, increases in age are correlated with increases in height. Then from, say, about age eighteen to the upper end of the age range, increases in age are no longer correlated with increases in height—generally after around age eighteen, people continue to get older but do not get taller. If we were to look at the two-dimensional chart, we could see the pattern and draw an inference: generally, increases in age correlate with increases in height but only until a certain age.

One application of machine learning is to ingest those million health records but also to look at many more than two variables at a time. For instance, the VA data includes records for more than six million patients, but each record contains many entries. If we were to imagine all the VA's data on a single spreadsheet, it would have twenty thousand columns and eighty billion rows.[16] Suppose someone were to write software that looked at all the variables in that data and asked which variables are correlated with heart disease, diabetes, or cancer. Because the machine learning software can ingest so much data, and because it can recognize patterns not just in two dimensions as you and I can, but across dozens or even hundreds of dimensions, the computer program could find correlations that even the best human research teams would miss.

In machine learning, developers can take a sample of the data— for instance, records of people who have already been diagnosed with heart disease, diabetes, or cancer—and use that sample to train an algorithm. Effectively, the algorithm learns what kinds of health records are associated with the diseases in question.

The second step in machine learning is to take this computer software model—this algorithm—that has been trained on the subset of health records and let it run on a much bigger data set, for example, the rest of the health records in the VA system. The algorithm might predict who is at a high risk for heart disease, diabetes, and cancer, and even tell us which variables are most closely correlated with these diseases. We might then have a sense for which variables we could change and therefore help prevent the diseases.[17]

This is just one example, but it helps show how AI, and specifically machine learning, can be used for good—to generate health-care benefits that humans on our own would be unable to produce. Machine learning is applicable not just in medicine but also in any application that can use historical data as training data.

If machine learning can make predictions in the medical field, it can do so on the battlefield, too. Now, instead of millions of health records, suppose we had millions of hours of Predator and Reaper video. Instead of data fields such as age, height, weight, and blood type, we had data fields with labels for the various kinds of things depicted in the video—a house, a vehicle, a person. If we had a big-enough training data set, we could conceivably teach a machine learning algorithm to predict whether the thing on the screen matches one of the sets of labeled items in the training data. For instance, the algorithm could use computer vision software to examine the features in the video, compare those features with what it learned from the labeled training data, and then tell us, for instance, that there's a 92 percent chance the thing on the screen is a house.

This was, according to reporting, the idea behind the Department of Defense's (DoD's) Project Maven. It is a machine learning application trained on video in which objects have already been labeled by humans. And then it attempts to identify objects from new video and fit the objects into the labeled categories.[18] The program made national news when Google engineers outside the program learned that Google had a contract to support the development of Project Maven. More than 3,100 engineers signed an open letter to Google CEO Sundar Pichai denouncing the company's involvement in "the business of war."[19]

For its part, the DoD has continuously maintained that it never intended for Maven to be used as an autonomous targeting system.

Instead, the program was meant only to reduce the workload on human intelligence analysts by automatically labeling objects on the screen. And still, says the DoD, no part of the weapon release process has been automated. In more practical terms, Maven sought to alleviate the massive amount of human labor behind the processing, exploitation, and dissemination of raw intelligence video.[20] Though we often think of Reaper operations requiring a pilot and sensor operator, they also require intelligence analysts to view the video feed— sometimes as many as three analysts to support a single aircraft. This demand for analysts means that just as a single twenty-four-hour aircraft sortie might be made up of three eight-hour shifts, requiring three pilots and three sensor operators, those same three shifts would also require as many as nine intelligence analysts. As the Predator and Reaper fleet grew, the personnel requirement to conduct the processing, exploitation, and dissemination became immense.[21] Project Maven was an effort to use machine learning to absorb some of that requirement for video analysis support to crews.

Image recognition, or *computer vision*, is not the only area in which AI will affect remote warfare. Engineers both inside and outside the DoD have already begun to take advantage of AI to enable aircraft to fly autonomously. Though any list I might generate here will undoubtedly be out-of-date by the time this book goes to print, here are a few noteworthy developments.

One important development of an autonomous airborne weapon, Israel's Harpy long-loiter antiradiation weapon, was introduced as early as 1990.[22] Instead of visually identifying targets, the Harpy is programmed to seek out electromagnetic spectrum emissions associated with adversary surface-to-air threat systems. The Harpy launches, proceeds to the target area, loiters overhead for hours, and waits for the right radio-frequency emissions. When it "hears" the right signal, rather than releasing a weapon, it becomes the weapon and dives toward the target.

There are also two notable British examples of AI weapons systems. The first, the Brimstone air-to-surface missile, has a laser-designator mode, like the Hellfire, but can also be used autonomously. In this mode, the missile flies to a designated area and uses a millimeter-wave radar to identify, track, and strike vehicles it finds there.[23] The Brimstone is so named because it is intended to replace the US-made Hellfire on British aircraft. Thus, the United States and United Kingdom can now rain down Hellfires and Brimstones on their adversaries—a poor choice in terminology. The British military is also developing an aircraft called Taranis. Though the program is described as an "advanced technology demonstrator" and not an operational aircraft, it aims to demonstrate technological capability for "marking targets, gathering intelligence, deterring adversaries, and carrying out strikes in hostile territory." At the same time, the aircraft's manufacturer boasts about the system's "full autonomy elements" and the fact that it is "under the control of a human operator."[24]

The United States is likewise developing autonomous systems wholly independent of the Reaper and Project Maven. In 2019, the air force established a program called Skyborg, which aims to produce the autonomous control system that will be embedded into low-cost autonomous aircraft that can operate in coordination with one another. Ultimately, the program intends to develop and apply autonomous technology not just to a single aircraft type but also to a family of unpiloted, autonomous systems. According to the Air Force Research Laboratory, these aircraft "will not replace human pilots. Instead, [they] will provide them with key data to support rapid, informed decisions."[25] The Skyborg program is just one example of a so-called loyal wingman, an autonomous drone that will fly alongside and support traditionally piloted aircraft. The central idea here is that the pilot of a more traditional aircraft, say, an F-35, might control a dozen relatively low-cost, autonomous wingmen. The human flight lead can decide which targets to engage, but the wingmen would autonomously maintain position in the formation, match weapons to targets, and maneuver around one another.[26]

In an entirely separate program, DoD researchers have attempted to automate dogfighting. In 2020, the Defense Advanced Research

Projects Agency held a competition among developers to build an autopilot system capable of defeating a human fighter pilot in a simulated dogfight.[27] There were important constraints, and a victory in the simulation doesn't translate directly to victory in the real world. Even so, it was startling to watch the live broadcast as the AI pilot won five of five air-to-air engagements against a well-trained human fighter pilot.

As the DoD and the US Air Force continue to incorporate AI and autonomy into their systems, doctrines, and concepts of operations, ethical questions will undoubtedly arise. The DoD recognized the ethical challenges early. In 2012, then Secretary of Defense Ash Carter published DoD Directive 3000.09 governing "the development and use of autonomous and semi-autonomous functions in weapons systems." The document requires that any system that employs autonomy in the use of force must be approved by the undersecretary of defense for policy.[28] The purpose of the document was not to prevent the military branches and agencies from developing autonomous systems but to ensure a sufficiently high level of oversight of autonomous systems that would be involved in the use of force.

Autonomy and AI are not synonyms. AI is a necessary condition for success in advanced autonomous systems. But the potential applications of AI are far broader than just autonomy. And the ethical concerns are broader too. In addition to the memorandum governing autonomous weapons, in 2020, the DoD adopted five ethical principles for AI. Any DoD applications using AI must be responsible, equitable, traceable, reliable, and governable.[29] While outlining these principles is an important step, translating them into actionable guidance for contracting, testing, and employment of AI systems will, I suspect, be an ongoing challenge.

With this landscape in view—developments in AI and autonomy and DoD attempts to impose ethical constraints on this new technology—what conclusions can we draw about the ethics of remote warfare as we move into the future? What influence does the marriage of AI and remote warfare have on the ethics of those systems? And, more centrally for this book, how will our lessons on the ethics of remote warfare remain relevant to AI-capable weapons systems?

I am not assuming a war between the United States and a peer adversary. As a citizen, a service member, and a military ethicist, I hope such a war between hegemonic states never comes. But as a service member and a military ethicist, I have a responsibility to think about what that kind of war might entail. First, we should expect remote warfare to become the norm rather than the exception. The bureaucracy that is the DoD has undergone a seismic shift in the last few years. The 2018 National Defense Strategy called for a transition from focusing on defending the United States and its interests against terrorist groups to defending against peer and near-peer adversary states. The states named in the National Defense Strategy are China, Russia, Iran, and North Korea, though clear emphasis is placed on China and Russia.[30] The DoD has responded to this call surprisingly quickly. One important implication of this shift is that the military can no longer assume that it will have freedom of movement throughout the theater of operations as it has had in Iraq and Afghanistan for two decades. The US Air Force can no longer assume that it will have air supremacy—that is, that it will have uncontested use of the air domain throughout the whole theater of operations. Finally, the United States cannot even assume that it will have uninhibited and unlimited use of the electromagnetic spectrum. Signals, perhaps even the radio-frequency signals used to control the Reaper, will likely be jammed, spoofed, or otherwise interfered with. The United States will have to contend with what strategists call A2/AD—anti-access, area-denial. Because adversaries will make it difficult for US assets to get close, American forces or, at the very least, some of them, will have to operate from further away—and that means new kinds of remote warfare.

We can imagine what some of these new types will look like. Instead of having the vast majority of the warfighting enterprise deployed forward into the combat environment, perhaps the majority of US warfighters will be distributed. Just as Predator and Reaper crews have flown their aircraft from bases around the United States, in the future, cyber warfare operators might engage the adversary from Fort Meade, Maryland, or from San Antonio, Texas. Perhaps bomber crews will use standoff weapons—AI-enabled, air-launched cruise missiles—rather than penetrating heavily defended enemy airspace. Or perhaps fighter pilots will remain at a safe distance while sending swarms of autonomous loyal wingmen, or drones, forward to conduct the air-to-air fighting.

The moral landscape of this new battlefield will look different from the battlefields of the past. But some of the moral elements that have been central to Predator and Reaper operations will remain relevant. In one sense, a war with a near-peer adversary would mark a return to some of the elements of the wars of the twentieth century. For instance, the air force chief of staff, General Charles Q. Brown, has attempted to brace his service for these changes. In a paper he released just after taking office in 2020, he described the transition: "Tomorrow's Airmen are more likely to fight in highly contested environments, and must be prepared to fight through combat attrition rates and risks to the Nation that are more akin to the era of the Second World War than the uncontested environment to which we have since become accustomed."[31] This reference to "attrition rates" is, on my reading, intended to startle a US Air Force that has enjoyed uncontested air supremacy for three decades. In one sense, his words suggest that aerial warfare of the future will impose significant physical risks on aircrews.

With these sorts of changes, we might think that there will be no more talk of riskless warfare or whether battles ought to be like duels. Perhaps we will just go back to the old way of war. But these interpretations misunderstand the broader context. Surely some aircrews will indeed have to face more risk. But as I have already suggested, operations will at the same time be more distributed. Even as fighter pilots expose themselves to enemy air threats, other service members, perhaps from various locations around the globe, will conduct operations in the cyber domain, in the electromagnetic spectrum, and in space operations to achieve effects alongside those pursued by the fighter pilot. Other pilots will remain outside the enemy's A2/AD threats and release long-range weapons into the theater of operations.

In this distributed environment, the virtues will be every bit as relevant to military practitioners as they have been to Predator and Reaper crews. Perhaps individual virtues required of military professionals might vary from case to case, but strength of character will remain a safeguard against the worst elements of war. As the United States moves toward semiautonomous systems and long-range weapons, there is a risk of a widening judgment gap. In the absence of a high-definition camera loitering overhead, will military practitioners still appreciate the magnitude of the destruction they cause? Will they be in a position to say no, to say that this particular strike under consideration violates the proportionality or necessity requirement? If so, it will only be because they have cultivated the virtues of moral courage and practical wisdom in advance.

We need not think only of hypothetical cases. In 2018, the Bashar al-Assad regime used chemical weapons against a town outside Damascus, Syria. The United States, the United Kingdom, and France responded with a coordinated attack involving 105 air- and ship-launched missiles against Syrian chemical weapons facilities. The DoD said that B-1 bombers released AGM-158 joint air-to-surface standoff missiles (JASSM).[32] According to some reports, this was the first use of the extended-range version of these missiles, the JASSM-ER. The original missile has a reported maximum range of about 230 miles, and its longer-range variant can be launched from more than 600 miles away.[33] In one sense, the three-nation attack on Syria looks like traditional warfare. Human aircrew members deployed to the US Central Command theater of operations, boarded their aircraft, and fired missiles at the enemy. But if the B-1 crews released their weapons from 200 or even 600 miles, did they really expose themselves to combat risk? Was their engagement with their adversaries anything like the depiction of the duel we have inherited from our medieval forebears? And if not, then have the lines between traditional warfare and remote warfare already begun to blur?

Just a year later, the National Defense Authorization Act established the US Space Force as an independent military branch. In one of that service's founding documents, General John W. Raymond, the first chief of space operations, asserts that the space force will provide a critical service "by operating in, from and to the space domain." He explains further: "The ability to legally transcend the most remote and protected national boundaries provides a unique opportunity to enable lethal and non-lethal effects against terrestrial targets."[34] As space operators seek to enable "lethal and non-lethal effects" from operations centers in the United States, we should again look to the implications for remote warfare. In a notional conflict with a peer adversary, surely some warfighters will accept great risk and proceed inside the enemy's threat ring. Others, though, will produce, or at the very least enable, effects from several thousand miles away as Predator and Reaper crews have done since 2001.

Many of the questions raised in previous chapters about the Predator and Reaper will be equally applicable in this new era of remote warfare. Can someone who faces no risk really be considered a combatant? Is taking the life of an enemy combatant from the other side of the world morally justified? And what character virtues do remote warfighters need to cultivate?

If military ethics is meant primarily as a constraint on the horrors of war, what will constrain the use of force in this future environment characterized by AI, autonomy, and distributed operations? In the future, if the horrors of war will be held back, if the destruction and death war brings with it are to be stemmed, it will be by means of the same constraints that have applied to war for centuries; practicing the tenets of *jus ad bellum* must prevent unjust wars from occurring—and when war is waged, practicing the tenets of *jus in bello* will ensure it is waged justly. I am not suggesting that just war and the study of ethics have performed these functions perfectly in the past; far from it. Instead, I am suggesting only that the restraints inherent in just war theory remain our best hope of preventing the calamities of war in the age of AI. In an era in which machines that learn will increasingly perform tasks we once thought suitable only for humans, we might take up the false belief that we can let our ethical guard down, or that we can trust the machines to act as our surrogates in the pursuit of the good. But this would be a mistake. In the looming age of AI and war, decision makers from the army private in the field to the combatant command leadership, and even to the elected and appointed civilians who lead and oversee the military will need more training and preparation in ethics, not less.

In the early years of the Predator's development, some US Air Force personnel mounted a crusade in the DoD for a view of the world according to which, with enough sensors and a sufficiently interconnected network, we could predict the future. *Effects-based operations* has come to mean many things to many people, but to some in the 1990s and early 2000s, it meant that if we got intelligence, surveillance, and reconnaissance right, we could not only see what was happening on every inch of the battlespace but also predict with near certainty what the adversary would do. We can still hear the echoes of that false promise in the way some in the DoD talk about AI. The Predator failed to deliver on that promise of a God's-eye view of the battlespace. Though we have yet to understand these tools fully, the limitations in AI and machine learning will likewise fail to deliver on that false promise.

In the opening pages of this book, I told the story of Lieutenant Dave Brown and the life-and-death decision he was asked to make in 2013. In the future, the context will change, and the technology will be different. The speed of operations will increase and so too, perhaps, will the consequences for failure. But I suspect that future wars will place moral demands on young military professionals similar to those placed on Dave Brown that night in 2013. Will we have done enough to prepare them to meet that challenge?

# ACKNOWLEDGMENTS

Though it is my name on the cover of this book, I am by no means the only person who contributed to the effort. My interest in publishing my work developed only because Flare Burdine encouraged me to publish that first paper nearly a decade ago. Since then, countless others have graciously read early (and often terrible) drafts of papers or carefully edited later ones. Jorge Garcia and Ken Himes supervised my masters' thesis on remotely piloted aircraft and martial virtue. I remain grateful for their patience and willingness to teach me. Later, my colleagues at the US Air Force Academy Department of Philosophy would indulge my ceaseless requests for feedback and editorial support. If memory serves, Claudia Hauer, Mike Growden, and Jim Cook bore the brunt of it—and I remain grateful to each of them. While in the doctoral program at Oxford, several professors and other graduate students were willing to engage with the arguments I was preparing for this book even though they fell well outside the scope of my dissertation. I am especially grateful to Cécile Fabre, Tom Simpson, Bob Underwood, Brianna Rosen, and Rhiannon Neilson. At the very nascent stages of this project, I had only a bad chapter draft or two and a small handful of sections that I didn't know how to pull together. In those moments, I turned to those whom I knew wouldn't turn me away: Dad, Mom, Katie, Kim,

Frank, Kenny, PAW, Dave, and Hammer, you suffered through some rubbish writing and poorly formed thoughts early on, and I knew you would because you're family. Thank you.

I also owe a debt of gratitude to a number of authors who have written on the ethics of drones. As is typical, I suspect, of books that make an argument, I disagree on some level with all the authors who have come before me. In the early mornings and late nights during which I wrote this book, those disagreements sometimes provided the motivation I needed to press on. I'll not list those authors here; I will say only that I have met several of them in person and they have proved to be, without exception, delightful people. To that august group, if you choose to read this book, though you will undoubtedly disagree, I hope only that you will see how much my thinking on this subject was in one way or another shaped by yours.

As I conducted research for the book, numerous air force leaders were supportive. Here, I thank, especially, Knuckles Campo, Travis Norton, and Eric Schmidt. I am especially indebted to the many Predator and Reaper crew members who were willing to speak with me about their own experiences and their own understanding of how ethics applies in their work. I should mention here that it did not seem right to me to make money on a book that I was able to write, in part, because of my own experience and relationships in the US Air Force remotely piloted aircraft community—it is a book on ethics, after all. Proceeds from the advance and any royalties from book sales will be donated to several charities, beginning with the Colorado Wounded Vet Run, REACT DC, and the National Multiple Sclerosis Society.

As I approached the final stages of this lengthy process, I waded into what were, for me, unfamiliar waters. I am grateful to Michael Mungiello at InkWell for his faithful and relentless representation of this book, and to Annie Trainque, without whom I'd never have secured Michael's support. And like falling dominoes, had it not been for Annie, I'd not have met Michael, and without Michael, I'd not have landed with PublicAffairs and with my editor, John Mahaney.

This book is far better than it would have been without the three of you. Thank you.

I have saved penultimate thanks for my wife, Megan. Parenting is honest work. There were early mornings, and weekends, and that week-long—what shall we call it; a writing retreat?—during which, while I was doing more writing, Megan was doing more parenting. Megan, this book would not have come to be without you and the sacrifices you made along the way. I am grateful.

I did say "penultimate" before. I know that God blesses me primarily through other people. Nowhere is that more evident than in my relationship with my wife, but the theme runs right through this long list of supporters, readers, editors, and interlocutors. I am at once grateful to those mentioned above and at the same time grateful to the God who gives good gifts to his children.

# NOTES

## INTRODUCTION

1. While most crewmembers quoted in the book are referred to by rank and first name, Dave Brown is a pseudonym. This decision was based on a prior arrangement with this pilot for a previous paper that is mentioned this story. See Chapa and DeWees, "Developing Character at the Frontier of Human Knowledge," *The Journal of Character and Leadership Integration* (Winter, 2016).

2. Just war theorists disagree on whether the proportionality principle is grounded in the expected military value to be achieved or the moral good to be achieved. Proportionality in just war theory is different from proportionality in the laws of war; a point to which I will return in a later chapter.

3. The term *airman* is the official term for all members of the US Air Force and not just men. Air force instructions typically refer to "All Airmen—officer, enlisted, and civilian." See, for example, USAF, *Air Force Instruction 36-2643: Air Force Mentoring Program* (Washington, DC: USAF, 2019), 1.

4. Alan Yuhas, "Airstrike That Killed Suleimani Also Killed Powerful Iraqi Militia Leader," *New York Times*, 3 January 2020; Matthew S. Schwartz, "Who Was the Iraqi Commander Also Killed in the Baghdad Drone Strike?," *NPR*, 4 January 2020.

5. Charlie Savage and Eric Schmitt, "Biden Secretly Limits Counterterrorism Drone Strikes Away from War Zones," *New York Times*, 3 March 2021.

6. Eric Schmitt, "A Botched Drone Strike in Kabul Started with the Wrong Car," *New York Times*, 21 September 2021.

7. For Biden and Secretary of Defense Austin references to "over-the-horizon" operations, see White House, "Remarks by President Biden on the Drawdown of U.S. Forces in Afghanistan," 8 July 2021, www.whitehouse

.gov/briefing-room/speeches-remarks/2021/07/08/remarks-by-president
-biden-on-the-drawdown-of-u-s-forces-in-afghanistan; US Department of Defense, "Secretary of Defense Lloyd J. Austin III Remarks
Before the Senate Armed Services Committee (as Prepared)," 28
September 2021, www.defense.gov/News/Speeches/Speech/Article/2791954
/secretary-of-defense-lloyd-j-austin-iii-remarks-before-the-senate-armed
-service. For commentators' interpretation of "over-the-horizon" as a reference
to remote warfare, see, for example, James Hohmann, "The Fantasy and Folly
of Trying to Look 'over the Horizon' in Afghanistan," *Washington Post*, 29
September 2021; Sarah Kreps and Paul Lushenko, "US Faces Immense Obsta-
cles to Continued Drone War in Afghanistan," *Tech Stream*, 14 October 2021.

8. Peter Bergen et al., *World of Drones: Examining the Proliferation, Development, and Use of Armed Drones* (Washington, DC: New America, 2017; updated 2019).

9. W. J. Hennigan, "Experts Say Drones Pose a National Security Threat—
and We Aren't Ready," *Time*, 31 May 2018.

10. Edoardo Maggio, "Putin Believes That Whatever Country Has the Best
AI Will Be 'the Ruler of the World,'" *Business Insider*, 4 September 2017.

11. Huw Roberts et al., "The Chinese Approach to Artificial Intelligence:
An Analysis of Policy, Ethics, and Regulation," *AI & Society* 36, no. 1 (2021).

12. US Department of Defense, "Summary of the 2018 Department of Defense Artificial Intelligence Strategy," Washington, DC, 8 November 2018, 5.

13. I return to questions about AI in Chapter 7.

14. Richard Whittle, *Predator: The Secret Origins of the Drone Revolution*
(New York: Henry Holt and Company, 2014), 156; Caitlin Lee, "The Culture of US Air Force Innovation: A Historical Case Study of the Predator Program," PhD diss., King's College, London, 2016, 198.

15. Walter J. Boyne, *Beyond the Wild Blue: A History of the Us Air Force,
1947–2007* (New York: Macmillan, 2007), 114.

16. Brian Glyn Williams, *Counter Jihad: America's Military Experience in
Afghanistan, Iraq, and Syria* (Philadelphia: University of Pennsylvania Press,
2017), 29.

17. Thomas H. Kean et al., *The 9/11 Commission Report*, National Commission on Terrorist Attacks upon the United States, 2004, 116–117.

18. Similar formulations refer to "riskless war," "extreme risk asymmetry,"
and the closely related claim that in this kind of warfare, "combat" is replaced
by "the hunt." See, for example, Medea Benjamin, *Drone Warfare: Killing
by Remote Control*, updated ed. (London: Verso, 2013); Christian Enemark,
*Armed Drones and the Ethics of War: Military Virtue in a Post-Heroic Age*, War,
Conflict and Ethics (London: Routledge, 2014), 81; John J. Kaag and Sarah
E. Kreps, *Drone Warfare* (Cambridge: Polity, 2014), 57; Grégoire Chamayou,
*A Theory of the Drone*, trans. Janet Lloyd (New York: New Press, 2013), 163.

19. US Air Force, *Annex 3-60: Targeting* (Maxwell Air Force Base, AL: Le-May Center for Doctrine Development and Education, 2019); US Air Force *Annex 3-03: Counterland Operations* (Maxwell Air Force Base, AL: LeMay Center for Doctrine Development and Education, 2019); Joint Chiefs of Staff, *Counterterrorism*, Joint Publication 3–26 (Washington, DC: Joint Staff, 2013).

20. US Air Force, Creech Air Force Base, "History of Creech Air Force Base," US Air Force, fact sheet, 16 May 2013, www.creech.af.mil/About-Us/Fact-Sheets/Display/Article/449127/history-of-creech-air-force-base.

21. In recent decades, a schism has developed among just war theorists. The question of whether war provides soldiers with a different set of moral justifications for killing than those available outside of war is nuanced and lies outside the scope of this book.

22. See Linda R. Robertson, *The Dream of Civilized Warfare: World War I Flying Aces and the American Imagination* (Minneapolis: University of Minnesota Press, 2003); Timothy P. Schultz, *The Problem with Pilots: How Physicians, Engineers, and Airpower Enthusiasts Redefined Flight* (Baltimore: Johns Hopkins University Press, 2018).

23. Frank Brennan, "The Lonely Impulse of Delight," *Journal of Palliative Care* 28, no. 4 (2012): 299.

24. H. H. Arnold and Ira Eaker, *This Flying Game* (New York: Funk & Wagnalls, 1936), 30.

25. C. J. Chivers, *The Fighters: Americans in Combat in Afghanistan and Iraq* (New York: Simon and Schuster, 2018), 302–303.

26. Most Western militaries recognize three levels of war: strategic, operational, and tactical. See, for example, Joint Chiefs of Staff, *Doctrine for the Armed Forces of the United States*, Joint Publication 1 (Washington, DC: Joint Staff, 2013); Warfare Branch Editor, *Land Operations* (Warminster, UK: Land Warfare Development Centre, 2017).

27. This is often referred to as the *moral hazard argument*. Though this argument appears throughout the literature on the ethics of remote weapons, its strongest formulation is in John Kaag and Sarah Kreps, "The Moral Hazard of Drones," *New York Times*, 22 July 2012; and Kaag and Kreps, *Drone Warfare*, 107. See also, Chamayou, *Theory of the Drone*, 188–191; Micah Zenko and Sarah Kreps, *Limiting Armed Drone Proliferation* (New York: Council on Foreign Relations, 2014); Hugh Gusterson, *Drone: Remote Control Warfare* (Cambridge, MA: MIT Press, 2015), 140–142; Rosa Brooks, *How Everything Became War and the Military Became Everything: Tales from the Pentagon* (New York: Simon & Schuster, 2016), 111.

28. See, for example, Chamayou, *Theory of the Drone*, 58; Kaag and Kreps, *Drone Warfare*, 2; Enemark, *Armed Drones and the Ethics of War*, 19–37; Gusterson, *Drone*, 15–21.

29. See, for example, Bergen et al., *World of Drones*; Elisa Catalano Ewers et al., "Drone Proliferation: Policy Choices for the Trump Administration," Center for a New American Security, Washington, DC, 1 June 2017; Zenko and Kreps, *Limiting Armed Drone Proliferation*.

30. This, among other things, is argued by Bradley Jay Strawser, "Moral Predators: The Duty to Employ Uninhabited Aerial Vehicles," *Journal of Military Ethics* 9, no. 4 (2010).

31. I, too, have argued for greater transparency in the US Air Force's remotely piloted aircraft operations. Joseph O. Chapa, "Remotely Piloted Aircraft and War in the Public Relations Domain," *Air & Space Power Journal* 28, no. 5 (2014).

32. Max Fisher, "Obama Finds Predator Drones Hilarious," *Atlantic*, 3 May 2010.

33. George W. Bush, "Selected Speeches of President George W. Bush," National Archives and Records Administration, Washington DC, 2008, 95.

34. See, respectively, Richard B. Cheney and Liz Cheney, *In My Time: A Personal and Political Memoir* (New York, London: Threshold Editions, 2011), 6; Winston Churchill, "The Few: Churchill's Speech to the House of Commons," Churchill Society, London, www.churchill-society-london.org.uk /thefew.html; Ronald Reagan, "Transcript of Address by Reagan on Libya," *New York Times*, 15 April 1986; Lyndon B. Johnson, "Address on Vietnam Before the National Legislative Conference, San Antonio, Texas," American Presidency Project, 29 September 1967, www.presidency.ucsb.edu/documents /address-vietnam-before-the-national-legislative-conference-san-antonio-texas; Franklin D. Roosevelt, "Speech by Franklin D. Roosevelt, New York (Transcript)," Address to Congress, 8 December 1941, Library of Congress.

## 1   THE PREDATOR PARADIGM

1. They are a combination of the horizontal stabilizer and elevator, thus, *stabilator*.

2. United Press International, "Air Force Finds Mechanical Failure Led to Crashes of Flying Team," *New York Times*, 11 April 1982; United Press International, "4 Pilots Killed as Stunt Planes Crash in Desert," *New York Times*, 19 January 1982.

3. Jacob R. McCarthy, "Tried and True to Air Force Blue: A Leader Remembered," Nellis Air Force Base, news release, 30 August 2007, www.nellis .af.mil/News/Article/285977/tried-and-true-to-air-force-blue-a-leader -remembered; Walter J. Boyne, "Creech," *Air Force Magazine*, 1 March 2005.

4. "Retired Gen. Creech, 'Father of the Thunderbirds,' Dies," *Las Vegas Sun*, 28 August 2003.

5. David A. Mindell, *Our Robots, Ourselves: Robotics and the Myths of Autonomy* (New York: Viking Adult, 2015), 120. Likewise, John Kaag and Sarah Kreps say that the Predator caused "a transformation of how the United States uses military force." Gusterson says, "Most people by now have a picture in their mind's eye of the drones themselves," and he describes the Predator's distinctive features. And though Ulrike Franke points to the ongoing debate over whether the Predator's impact on warfare is sufficient to warrant the attention it has received, she nevertheless admits that it "has become the image of drone warfare." Kaag and Kreps, *Drone Warfare*, 5; Gusterson, *Drone*, 1–2; Ulrike Franke, "The Unmanned Revolution: How Drones Are Revolutionising Warfare," PhD diss., University of Oxford, 2018, 47.

6. As Caitlin Lee puts it, Predator's development proceeded in "fits and starts." In 2001, just as the Predator paradigm was about to begin, Thomas Ehrhard, in his foundational work on the development of unmanned aerial vehicles, wrote, "Even with the rapid advancement of information and flight technologies in the current era, unmanned aerial vehicles still remain more of an oddity than a serious combat capability." Lee, "Case Study of the Predator," 5. Thomas Ehrhard, "Unmanned Aerial Vehicles in the United States Armed Services: A Comparative Study of Weapon System Innovation," PhD diss., Johns Hopkins University, Baltimore, 2001, 651.

7. Ehrhard, "Unmanned Aerial Vehicles," 52.

8. This assumption is pervasive in the literature, even though it is seldom explicitly stated. I include in this category the claims—I think they are exaggerated claims—that the Predator is robotic or semiautonomous. For "push-button warfare," see Jesse Kirkpatrick, "Drones and the Martial Virtue Courage," *Journal of Military Ethics* 14, no. 3–4 (2015): 211; Jane Mayer, "The Predator War," *New Yorker* 26 (2009): 36–45. For "robotic," see David Kilcullen testimony in US House of Representatives, Committee on Armed Services, *Effective Counterinsurgency: The Future of the U.S.-Pakistan Military Partnership*, hearing held 23 April 2009 (Washington, DC: US Government Printing Office, 2010); Max Hastings, "Gatwick Drone Shambles Is Just a Taste of What's to Come," *Times* (London), 21 December 2018; Anton Petrenko, "Between Berserksgang and the Autonomous Weapons Systems," *Public Affairs Quarterly* 26, no. 2 (2012): 82; L. Royakkers and R. van Est, "The Cubicle Warrior: The Marionette of Digitalized Warfare," *Ethics and Information Technology* 12, no. 3 (2010): 289; M. Shane Riza, "Two-Dimensional Warfare: Combatants, Warriors, and Our Post-Predator Collective Experience," *Journal of Military Ethics* 13, no. 3 (2014): 259; James DeShaw Rae and John T. Crist, *Analyzing the Drone Debates: Targeted Killing, Remote Warfare, and Military Technology*, Palgrave Pivot (Basingstoke, UK: Palgrave Macmillan, 2014), 91; Rob Sparrow, "War Without Virtue?," in *Killing by Remote Control,*

ed. Bradley Jay Strawser (Oxford: Oxford University Press, 2013), 86; Kaag and Kreps, *Drone Warfare*. For "semiautonomous," see Daniel Brunstetter and Megan Braun, "The Implications of Drones on the Just War Tradition," *Ethics & International Affairs* 25, no. 03 (2011): 338; Kaag and Kreps, *Drone Warfare*, vii; Royakkers and van Est, "The Cubicle Warrior," 293.

9. For "hunter-killer," see Chamayou, *Theory of the Drone*, 34; Derek Gregory, "From a View to a Kill," *Theory, Culture & Society* 28, no. 7–8 (2012): 193; Neil C. Renic, "UAVs and the End of Heroism? Historicising the Ethical Challenge of Asymmetric Violence," *Journal of Military Ethics* 17, no. 4 (2018): 192.

10. Mindell, *Our Robots, Ourselves*, 120.

11. Scott Shane, *Objective Troy: A Terrorist, a President, and the Rise of the Drone* (New York: Seal Books, 2016), 73.

12. For the driving simulator study, see Adam Waytz, Joy Heafner, and Nicholas Epley, "The Mind in the Machine: Anthropomorphism Increases Trust in an Autonomous Vehicle," *Journal of Experimental Social Psychology* 52 (2014). For other applications, see, for example, Juliana Schroeder and Matthew Schroeder, "Trusting in Machines: How Mode of Interaction Affects Willingness to Share Personal Information with Machines," Proceedings of the 51st Hawaii International Conference on System Sciences, in *Social and Psychological Perspectives in Collaboration Research*, 3 January 2018; Ewart J. de Visser et al., "Almost Human: Anthropomorphism Increases Trust Resilience in Cognitive Agents," *Journal of Experimental Psychology: Applied* 22, no. 3 (2016); Maha Salem et al., "Effects of Gesture on the Perception of Psychological Anthropomorphism: A Case Study with a Humanoid Robot," paper presented at the International Conference on Social Robotics, Berlin, 2011.

13. For example, the C-17 Globemaster III. See US Air Force, *Air Force Instruction 11-2C-17*, Vol. 3, *Operations Procedures* (Washington, DC: USAF, 2015), 73; Bundesstelle für Flugunfalluntersuchung (German Federal Bureau of Aircraft Accident Investigation), *Investigation Report: BFU EX010-11* (Braunschweig, Germany: Bundesstelle für Flugunfalluntersuchung, 2011), www.bfu-web.de/EN/Publications/Investigation%20Report/2011/Report_11_EX010_B777_Munic.pdf?__blob=publicationFile.

14. For example, the RQ-4 Global Hawk. See US Department of Defense, *Unmanned Aerial Vehicles Roadmap* (Washington, DC: DoD, 2000), 54.

15. Schultz, *The Problem with Pilots*, 161. For "morale stick," see also Mindell, *Our Robots, Ourselves*, 117–118.

16. Though the Predator's technological history has been told in numerous publications, no one I know of has provided a more thorough account of the cultural developments than Caitlin Lee, PhD. In this section, I cite her work frequently.

17. Lee, "Case Study of the Predator," 115.

18. At various times and in various pockets of the air force Predator and Reaper community, the control interface has been called both a *cockpit* and a *ground control station*. For consistency, I call it a cockpit throughout the book.

19. Lee, "Case Study of the Predator," 116.

20. Ibid., 153–155.

21. Ibid., 179.

22. Thomas Ehrhard writes, "Army operators at Fort Huachuca were crashing [Hunters] at an alarming rate." Ehrhard, "Unmanned Aerial Vehicles," 540.

23. Ehrhard, *Air Force UAVs: The Secret History*, ed. David A. Deptula (Arlington, VA: Mitchell Institute Press, 2010), 51.

24. For instance, though General McPeak saw the Predator as a threat to the cultural supremacy of fighter pilots in the US Air Force, years later, he admitted that he would have pursued it if he had thought the army would have turned it into a successful program. "If I thought the Army was going to make something out of it, I would have told them it was ours. . . . This is airpower! You're poaching on my territory!" Lee, "Case Study of the Predator," 145, 81.

25. Ehrhard, *Air Force UAVs*, 51. This policy changed to a limited degree over time. At various times, officers with an aeronautical rating of navigator or combat systems officer, along with a Federal Aviation Administration pilot's license, have also been permitted to fly the Predator and Reaper.

26. Lee, "Case Study of the Predator," 155–156.

27. US Air Force, *Air Force Instruction 11-202*, Vol. 3, *General Flight Rules* (Washington, DC: USAF, 2019), 7.

28. Major Bryce, interview with author, Creech Air Force Base (AFB), NV, 18 March 2019. For reasons of privacy and security, many of my interviewees requested that I use their first names only.

29. Squadron Leader Russ, interview with author, 22 March 2019, Creech AFB, NV.

30. I deal with this question directly in Chapter 2, "Riskless Warfare?"

31. Hanson W. Baldwin, "The 'Drone': Portent of Push-Button War; Recent Operations Point to the Pilotless Plane as a Formidable Weapon in War's New Armory," *New York Times*, 25 August 1946. Baldwin presciently predicted that "pilotless planes flown by remote control . . . will soon be a familiar sight in the skyways of the world." I discovered this article in Franke, "The Unmanned Revolution."

32. The aphorism comes from Carl H. Builder, *The Army in the Strategic Planning Process: Who Shall Bell the Cat?* (Santa Monica, CA: RAND, 1987), 26, www.rand.org/pubs/reports/R3513.html#download.

33. Paula G. Thornhill, *"Over Not Through": The Search for a Strong, Unified Culture for America's Airmen* (Santa Monica: RAND, 2012), 8; US Air Force,

*Annex 3-0: Operations and Planning* (Maxwell AFB, AL: LeMay Center for Doctrine Development and Education, 2016).

34. See Thornhill, *"Over Not Through."*

35. Donald L. Miller, *Masters of the Air: America's Bomber Boys Who Fought the Air War Against Nazi Germany* (New York: Simon & Schuster, 2006), 202.

36. Cathal J. Nolan, *The Allure of Battle: A History of How Wars Have Been Won and Lost* (New York: Oxford University Press, 2019), 144; also 51, 92.

37. Whittle, *Predator*, 73.

38. For "sensor to shooter," see Benjamin S. Lambeth, "Airpower, Spacepower, and Cyberpower," *Joint Force Quarterly* 60, no. 1 (2011): 47; Benjamin S. Lambeth, *NATO's Air War for Kosovo: A Strategic and Operational Assessment* (Santa Monica: RAND, 2001), 143; Benjamin S. Lambeth, *Air Power Against Terror: America's Conduct of Operation Enduring Freedom* (Santa Monica: RAND, 2005), 342, 50; Frank Sauer and Niklas Schörnig, "Killer Drones: The 'Silver Bullet' of Democratic Warfare?," *Security Dialogue* 43, no. 4 (2012): 370; Gregory Wilson, "A Time-Critical Targeting Roadmap," thesis, Air Command and Staff College, Maxwell AFB, AL, 2002.

39. Lee, "Case Study of the Predator," 129.

40. Whittle, *Predator*, 130–132; Lee, "Case Study of the Predator," 184–189.

41. Jonathan L. Jackson, "Solving the Problem of Time-Sensitive Targeting," paper, Naval War College, 2003, 2, 5; Wilson, "Time-Critical Targeting Roadmap," 3–8; John P. McDonnell, "Apportion or Divert? The JFC's Dilemma: Asset Availability for Time-Sensitive Targeting," thesis, Naval War College, 2002, 4.

42. Whittle, *Predator*, 147–148.

43. See, for example, ibid., 214–228; Houston R. Cantwell, "Beyond Butterflies: MQ-1 Predator and the Evolution of Unmanned Aerial Vehicles in Air Force Culture," thesis, School of Advanced Air and Space Studies, 2007, 61; Bernard D. Rostker et al., *Building Toward an Unmanned Aircraft System Training Strategy* (Santa Monica, CA: RAND, 2014), 35; Mindell, *Our Robots, Ourselves*, 143–148.

44. For the casualty numbers, see Princeton N. Lyman and F. Stephen Morrison, "The Terrorist Threat in Africa," *Foreign Affairs* 83, no. 1 (January–February 2004): 75. For the assignment of responsibility to al Qaeda, see Thomas H. Kean et al., *The 9/11 Commission Report*, National Commission on Terrorist Attacks upon the United States (2004), 115–116.

45. Chris Woods, *Sudden Justice: America's Secret Drone Wars* (New York: Oxford University Press, 2015), 39.

46. Lee, "Case Study of the Predator," 78–79.

47. The reduced risk to aircrew is crucial to the debate over the ethics of remote weapons. I address questions of risk directly in Chapter 2, "Riskless Warfare?"

48. Fogleman's claim was about the ability to hit targets. There is a closely related question about which objects or people ought to be considered targets. Many have argued that some air force senior leaders in the 1990s elevated the concept of effects-based operations to such omniscience that it had deleterious effects in the years that followed. See, for example, John T. Correll, "The Assault on EBO: The Cardinal Sin of Effects-Based Operations Was That It Threatened the Traditional Way of War," *Air Force Magazine*, January 2013, 50–54; James N. Mattis, "USJFCOM Commander's Guidance for Effects-Based Operations," *Joint Force Quarterly* 4, no. 51 (2008); Andrew Cockburn, *Kill Chain: The Rise of the High-Tech Assassins* (London: Verso, 2015).

49. Nolan calls these wars of "grinding exhaustion." Nolan, *The Allure of Battle*, 9.

50. Mindell, *Our Robots, Ourselves*, 132; James Holland, *The Battle of Britain: Five Months That Changed History, May–October 1940* (London: Corgi, 2011), 698. Robertson, *Dream of Civilized Warfare*, 92.

51. Richard P. Hallion, "Fighter Aircraft," in *The Oxford Companion to American Military History*, ed. Richard Holmes, Charles Singleton, and Spencer Jones (Oxford: Oxford University Press, 2000).

52. See, for example, R. Michael Worden, *Rise of the Fighter Generals: The Problem of Air Force Leadership, 1945–1982* (Maxwell AFB, AL: Air University Press, 1998).

53. US Air Force, *Annex 3-60: Targeting*; US Air Force, *Annex 3-03: Counterland Operations*; Joint Chiefs of Staff, *Counterterrorism*, Joint Publication 3–26.

54. For variations on this claim, see Elke Schwarz, "Written Submission of Evidence to the All Party Parliamentary Group (APPG) on Drones: Ethical Challenges," All Party Parliamentary Group on Drones, December 2017, http://appgdrones.org.uk/wp-content/uploads/2014/08/10.-Elke-APPG -Drones-041217.pdf, 4; Sarah E. Kreps, *Drones: What Everyone Needs to Know* (New York: Oxford University Press, 2016); Jamie Shea, "Precision Strike Capabilities: Political and Strategic Consequences," preface to *Precision Strike Warfare and International Intervention: Strategic, Ethico-Legal, and Decisional Implications*, ed. Mike Aaronson et al. (London: Routledge, 2015), xxii; Conway Waddington, "Drones: Degrading Moral Thresholds for the Use of Force and the Calculations of Proportionality," in *Precision Strike Warfare and International Intervention: Strategic, Ethico-Legal, and Decisional Implications*, ed. Mike Aaronson et al. (London: Routledge, 2015), 124; Christopher Coker, "Ethics of Counterinsurgency," in *The Routledge Handbook of Insurgency and Counterinsurgency*, ed. Paul B. Rich and Isabelle Duyvesteyn (London: Routledge, 2012), 124.

55. Takur Ghar is now often referred to as Roberts' Ridge, after Neil Roberts, the first American casualty in the engagement.

56. Linda D. Kozaryn, "Fog, Friction Rule Takur Ghar Battle," American Forces Press Service, news release, 24 May 2002, https://archive.defense.gov/news/newsarticle.aspx?id=44019.

57. Dan Schilling and Lori Chapman Longfritz, *Alone at Dawn* (New York: Grand Central Publishing, 2019), 303; Sean Naylor, *Not a Good Day to Die: The Untold Story of Operation Anaconda*, rev. ed. (London: Penguin, 2006).

58. Schilling and Chapman Longfritz, *Alone at Dawn*.

59. Paul Szoldra, "Watch John Chapman's Incredible Heroics in the First Medal of Honor Action Ever Recorded on Video," *Task and Purpose* (2019).

60. Director Sam Hargrave is currently developing a film about Chapman's life and final act of heroism starring Jake Gyllenhaal. James Barber, "Jake Gyllenhaal Set to Play Medal of Honor Recipient John Chapman in 'Combat Control,'" *Military.com*, 22 March 2021.

61. Sean Naylor, *Not a Good Day to Die: The Untold Story of Operation Anaconda*, rev. ed. (London: Penguin, 2006), 356.

62. Whittle, *Predator*, 278–298.

63. James Thompson, "Phantom of Takur Ghar: The Predator above Roberts Ridge," news release, 30 August 2018, www.acc.af.mil/News/Article-Display/Article/1617739/phantom-of-takur-ghar-the-predator-above-roberts-ridge/.

64. The quote is attributed to Colonel Ed Boyle in Whittle, *Predator*, 298.

65. Richard Pallardy, "2010 Haiti Earthquake," in *Encyclopædia Britannica* (Chicago: Encyclopædia Britannica, 2020).

66. Douglas M. Fraser and Wendell S. Hertzelle, "Haiti Relief: An International Effort Enabled Through Air, Space, and Cyberspace," *Air and Space Power Journal* 24, no. 4 (2010): 9.

67. Nathan Broshear, "Airmen Fly Predator in Controlled Airspace over Haiti," US Air Force, 12th Air Force Southern Public Affairs, 29 January 2010, www.af.mil/News/Article-Display/Article/117786/airmen-fly-predator-in-controlled-airspace-over-haiti.

68. Fraser and Hertzelle, "Haiti Relief," 9.

69. Bart Elias, *Unmanned Aircraft Operations in Domestic Airspace: U.S. Policy Perspectives and the Regulatory Landscape*, Congressional Research Service report 7-5700 (Washington, DC, 27 January 2016), www.hsdl.org/?view&did=789767, i.

70. See, for example, Oana Lungescu, Giampaolo Di Paola, and Charles Brouchard, "Press Briefing," North Atlantic Treaty Organization, news release, 31 March 2011, www.nato.int/cps/en/natolive/opinions_71897.htm; John Mueller, "The Perfect Enemy: Assessing the Gulf War," *Security Studies* 5,

no. 1 (1995); Will Skowronski, "Reapers and the RPA Resurgence: The MQ-9 Can Perform Strike, Coordination and Reconnaissance Against High-Value Targets," *Air Force Magazine*, August 2016.

71. Deborah C. Kidwell, "The U.S. Experience: Operational," in *Precision and Purpose: Airpower in the Libyan Civil War*, ed. Karl P. Meuller (Santa Monica, CA: RAND, 2015), 127.

72. Ibid., 146.

73. Major Bryce, interview with author, Creech AFB, NV, 18 March 2019.

74. For the preceding claims, see, respectively, Colin Clark, "Reaper Drones: The New Close Air Support Weapon," *Breaking Defense*, 10 May 2018; Christian Clausen, "Combat RPAs Integral in Defeating ISIS," US Air Force, 432nd Wing/432nd Air Expeditionary Wing Public Affairs, 5 December 2017; Valerie Insinna, "In the Fight against ISIS, Predators and Reapers Prove Close-Air Support Bona-Fides," *Defense News*, 8 August 2017; Christian Clausen, "Next Level of RPA Operations, USAFCENT Commander Recognizes Airmen," US Air Force, 432nd Wing/432nd Air Expeditionary Wing Public Affairs, 10 January 2018, www.af.mil/News/Article-Display/Article/1388206/combat-rpas-integral-in-defeating-isis.

75. Adawn Kelsey, "RPA Community Launches 65th CAP, Meets SecDef Initiative," press release, 5 June 2014, www.creech.af.mil/News/Article-Display/Article/670033/rpa-community-launches-65th-cap-meets-secdef-initiative.

76. The "odd bird" description is from Ehrhard, "Unmanned Aerial Vehicles," 52.

77. For "hunter-killer," see Chamayou, *Theory of the Drone*, 34; Gregory, "From a View to a Kill," 193; Renic, "UAVs and the End of Heroism?," 192.

78. For "Bugsplat," see Benjamin, *Drone Warfare*, 160; Gusterson, *Drone*, 38. For "collateral damage" as a euphemism, see Michael J. Boyle, "The Costs and Consequences of Drone Warfare," *International Affairs* 89, no. 1 (2013): 25n146; C. A. J. Coady, *Morality and Political Violence* (Cambridge: Cambridge University Press, 2008), 132; Laurie Calhoun, "Drone Killing and the Disastrous Doctrine of Double Effect," *Peace Review* 27, no. 4 (2015): 444.

79. Lee, "Case Study of the Predator," 113.

80. John Schlight, "Project CHECO Southeast Asia Report: Jet Forward Air Controllers in SEAsia," report, Hickam AFB, HI, 15 October 1969, 29; Dennis Larm, "Expendable Remotely Piloted Vehicles for Strategic Offensive Airpower Roles," MPhil thesis, School of Advanced Airpower Studies (Maxwell AFB, AL, 1996), 31; Arthur J. C. Lavalle, ed., *The Tale of Two Bridges and the Battle for the Skies over North Vietnam* (Columbia, PA: Diane Publishing, 1976), 1:152–153; James E. Moschgat, *The Composite Wing: Back to the Future* (Maxwell AFB, AL: Air University Press, 1992), 68.

81. US Army Aviation and Missile Life Cycle Management Command (AMCOM), "Hellfire," AMCOM, information page, n.d., https://history .redstone.army.mil/miss-hellfire.html.

## 2   RISKLESS WARFARE?

1. Alfred Asch, Hugh R. Graff, and Thomas A. Ramey, *The Story of the Four Hundred and Fifty-Fifth Bombardment Group (H) WWII: Flight of the Vulgar Vultures* (Appleton, WI: Graphic Communications Center, 1991), 84–85, 117.

2. See, for example, Enemark, *Armed Drones and the Ethics of War*, 7, 96; Chamayou, *Theory of the Drone*, 17; Marcus Schulzke, "Rethinking Military Virtue Ethics in an Age of Unmanned Weapons," *Journal of Military Ethics* 15, no. 3 (2016): 194.

3. Some military history scholars use the term *warrior ethos* to distinguish between the ethos of modern, professionalized soldiers and the warrior age that preceded it. This usage has a negative connotation. Warriors, in this usage, aren't constrained by conceptions of the laws of war or military ethics. On the contrary, most military ethics scholars use the term to refer to a constraining force. The warrior ethos is that which enables warriors to keep their humanity even while committing acts of violence. For *warrior ethos* with a negative connotation, see, for example, Roger Wertheimer, *Empowering Our Military Conscience: Transforming Just War Theory and Military Moral Education*, Military and Defence Ethics (Farnham, UK: Ashgate, 2010), 139; Ryan Noordally, "On the Toxicity of the 'Warrior' Ethos," *Wavell Room: Contemporary British Military Thought*, 28 April 2020. Cathal Nolan, *The Allure of Battle*, does not comment directly on the warrior ethos but does seem to use the term *warriors* to refer to medieval soldiers-for-hire in contrast to the professionalized militaries of the modern period and beyond.

4. See, for example, Shannon E. French, *The Code of the Warrior: Exploring Warrior Values Past and Present*, 2nd ed. (Lanham, MD: Rowman & Littlefield, 2017); Christopher Coker, *The Warrior Ethos: Military Culture and the War on Terror*, LSE International Studies (New York; London: Routledge, 2007); Pauline M. Kaurin, *The Warrior, Military Ethics and Contemporary Warfare: Achilles Goes Asymmetrical*, Military and Defence Ethics (Farnham, UK: Ashgate, 2014).

5. Find the following terms and their association with the warrior ethos with their respective references. "Heroism," Coker, *The Warrior Ethos*, 3–4; Chamayou, *Theory of the Drone*, 98. "Cult of youthful heroism," Victor Davis Hanson, *The Western Way of War: Infantry Battle in Classical Greece* (New York: A. A. Knopf, 1989), 225. "Courage," Enemark, *Armed Drones and the Ethics of War*, 6, 117; Coker, *The Warrior Ethos*, 3, 34. "Risk-taker," Enemark, *Armed Drones and the Ethics of War*, 6. "Emotional connection," M. Shane

Riza, *Killing Without Heart: Limits on Robotic Warfare in an Age of Persistent Conflict* (Washington, DC: Potomac Books, 2013), 270. "Existential," Coker, *The Warrior Ethos*, 5–6, 29, 54, 77; Riza, "Two-Dimensional Warfare," 269. "What it means to go to war," "to be a soldier," and "to be a warrior," Riza, *Killing Without Heart*, 5; Enemark, *Armed Drones and the Ethics of War*, 63; Riza, "Two-Dimensional Warfare," 269. "Transformative" or "transformation," Coker, *The Warrior Ethos*, 4–5. "Sacrifice," Coker, *The Warrior Ethos*, 5, 61, 71; Chamayou, *Theory of the Drone*, 17, 98.

6. Stephen Budiansky, *Oliver Wendell Holmes: A Life in War, Law, and Ideas* (New York: W. W. Norton & Company, 2019), 90.

7. Ibid., 93.

8. Oliver Wendell Holmes, *Speeches*, Making of Modern Law (Boston: Little, Brown, 1918), 62.

9. Coker calls this "the warrior myth." Coker, *The Warrior Ethos*, 34.

10. For instance, according to Riza, warriors are dedicated to "a personal growth through the profession of arms." For them, "combat is the ultimate and artistic expression" of their lives. Warriors see combat as an "intricate dance, a test of personal will and technical skill." Enemark likewise extols the "professional virtue of the warrior as a courageous risk-taker." Riza, "Two-Dimensional Warfare," 260–261, 68–69; Enemark, *Armed Drones and the Ethics of War*, 6.

11. Alexander Hamilton, "Letter to Edward, 11 November 1769," *The Papers of Alexander Hamilton: Digital Edition*, vol. 1, *1768–1778* (Charlottesville: University of Virginia Press, Rotunda, 2011), http://rotunda.upress.virginia .edu/founders/default.xqy?keys=ARHN-print-01-01-02-0002 See also Ron Chernow, *Alexander Hamilton* (New York: Penguin, 2004), 18–19, 30–31.

12. William V. Harris, *War and Imperialism in Republican Rome: 327–70 B.C.* (Oxford and New York: Clarendon Press and Oxford University Press, 1979), 17.

13. Hoplites were heavily armed infantry soldiers of ancient Greece. Hanson, *Western Way of War*, 220.

14. J. Glenn Gray, *The Warriors: Reflections on Men in Battle* (Lincoln: University of Nebraska Press, 1998), 216.

15. The origin of the phrase is contested. John D. Wright says that it began when farmers saw elephants for the first time at a traveling circus. Dan Hampton says that the *elephant* in the phrase refers to Hannibal's crossing of the Alps, but he does not say when the phrase was adopted. John D. Wright, *The Language of the Civil War* (Westport, CT: Oryx Press, 2001), 265; Dan Hampton, *Viper Pilot: The Autobiography of One of America's Most Decorated F-16 Combat Pilots* (New York: William Morrow, 2012).

16. Ulysses S. Grant, *The Papers of Ulysses S. Grant: Volume 1: 1837–1861*, vol. 1 (Carbondale: Southern Illinois University Press, 1967), 105.

17. Arthur Clutton-Brock and André Chevrillon, *Letters from a Soldier of France 1914–1915: Wartime Letters from France* (South Yorkshire, UK: Pen and Sword, 2014), 34.

18. Gray, *The Warriors*, 51.

19. Coker, *The Warrior Ethos*, 5. Likewise, Pauline Shanks Kaurin says that "when it comes to thinking about war and warriors, first there was Achilles, and then the rest followed." Kaurin, *Warrior, Military Ethics*, 1. Peter Lee similarly suggests that "Achilles has been portrayed as the archetypal military hero throughout the Western history of war." Peter Lee, "Heroes and Cowards: Genealogy, Subjectivity and War in the Twenty-First Century," *Genealogy* 2, no. 2 (2018): 2.

20. C. J. Chivers, "A Changed Way of War in Afghanistan's Skies," *New York Times*, 15 January 2012; Sarah B. Sewall, *Chasing Success: Air Force Efforts to Reduce Civilian Harm* (Maxwell AFB, AL: Air University Press, 2016), 164, www.airuniversity.af.edu/Portals/10/AUPress/Books/B_0142_SEWALL _CHASING_SUCCESS.pdf.

21. For bias for action, see David H. Berger, "Commandant's Planning Guidance: 38th Commandant of the Marine Corps," Washington, DC, 2019, 16, www.marines.mil/Portals/1/Publications/Commandant's%20 Planning%20Guidance_2019.pdf. For close-combat lethality in complex terrain, see James N. Mattis, *Summary of the National Defense Strategy*, Department of Defense, Washington, DC, 2018, 6.

22. Coker, *The Warrior Ethos*, 7, 57.

23. Ibid., 8.

24. For "wrath," see Homer, *The Iliad of Homer*, trans. Ernest Myers, Walter Leaf, and Andrew Lang (New York: Modern Library, 1929), 1; Homer, *The Iliad of Homer*, trans. Alexander Pope, Chadwyck-Healey Literature Collections (Cambridge: Chadwyck-Healey, 1992), 1; Homer, *Homer's Iliad*, trans. George Chapman, MHRA Tudor & Stuart Translations (Cambridge: Modern Humanities Research Association, 2017), 32–33. Some translations refer to Achilles's "rage." For this claim and the claim that Achilles's wrath drives the book's plot, see French, *Code of the Warrior*, 38.

25. Plato, "The Republic," in *The Collected Dialogues of Plato*, ed. Edith Hamilton, Bollingen Series (Princeton, NJ: Princeton University Press, 2002), 71:636.

26. Coker, *The Warrior Ethos*, 65.

27. French, *Code of the Warrior*, xiv, 23, 35–36, 63. Coker, *The Warrior Ethos*, 64.

28. James L. Cook, "Review of *Killing Without Heart: Limits on Robotic Warfare in an Age of Persistent Conflict*, by Shane M. Riza," *Journal of Military Ethics* 13 (2014): 110.

29. Coker says, "What is most mythic of all is the willingness of the warrior to consecrate his life to the rest of us." Olsthoorn suggests that peace and security depend on "the willingness of some to make sacrifices for the security of others." Enemark rejects remote warfare because, among other things, remote warriors are not willing to risk their lives as traditional warriors are. Finally, Neil Renic says that "the warrior ethos has hitherto required . . . the presence of physical courage, as well as the willingness to risk harm." Coker, *The Warrior Ethos*, 39; Peter Olsthoorn, "Honor as a Motive for Making Sacrifices," *Journal of Military Ethics* 4, no. 3 (2005): 190; Enemark, *Armed Drones and the Ethics of War*, 108; Renic, "UAVs and the End of Heroism?," 193.

30. Stephen E. Ambrose, *Band of Brothers* (London: Simon and Schuster UK, 2017), 62.

31. Karl Marlantes, *What It Is Like to Go to War* (London: Corvus, 2011), 44. Emphasis in original.

32. Hanson, *Western Way of War*, 226.

33. Enemark is concerned that remote warfare is "enabling a form of violence so fundamentally different in nature that it does not count as war." Grégoire Chamayou argues that in the age of remote warfare, "the fundamental structure of this type of warfare is no longer that of a duel, of two fighters facing each other," but instead is that of hunter and prey. Hugh Gusterson applies the same analogy, saying that remote warfare challenges the duel-like nature of warfare: "By its nature, war is structured around the reciprocal infliction of pain and death. It is a contest, maybe an uneven one, but a contest nonetheless." Enemark, *Armed Drones and the Ethics of War*, 6, 59–60; Chamayou, *Theory of the Drone*, 33; Gusterson, *Drone*, 145.

34. Carl von Clausewitz, *On War*, trans. Michael Howard and Peter Paret (Oxford: Oxford University Press, 2008), 13.

35. For the "field of honor," see, for example, Chernow, *Alexander Hamilton*, 308.

36. Joyal Mark, Iain McDougall, and J. C. Yardley, *Greek and Roman Education: A Sourcebook* (London: Routledge, 2009), 15.

37. Chamayou, *Theory of the Drone*, 52.

38. Ibid.; Dave Blair and Karen House, "Avengers in Wrath: Moral Agency and Trauma Prevention for Remote Warriors," *Lawfare*, 12 November 2017.

39. Clausewitz, *On War*, 13. Emphasis added.

40. Ibid., 28.

41. For example, the US military's joint publication on doctrine includes references to Clausewitz's views on war as a "duel on a larger scale," "as an act of force to compel our enemy," as a "continuation of politics by other means," and as "a violent clash of wills." Joint Chiefs of Staff, *Doctrine for the Armed Forces*, I-3.

42. Ute Frevert, *Men of Honour: A Social and Cultural History of the Duel* (Cambridge: Polity, 1995), 11.

43. French, *Code of the Warrior*, 130.

44. See Frevert, *Men of Honour*, 171. For the transition from chivalric to modern codes, see for example, Olsthoorn, "Honor as a Motive," 187.

45. See, respectively, Max Hastings, *Overlord: D-Day and the Battle for Normandy* (London: Pan Books, 2015), 63, 172; Boyne, *Beyond the Wild Blue*, 162–163; Robert Coram, *Boyd: The Fighter Pilot Who Changed the Art of War* (Boston: Little, Brown, 2002), 424; Homer, *The Odyssey*, trans. Alexander Pope (Doylestown, PA: Wildside Press, 2003), 67–93.

46. Antulio Joseph Echevarria, *Clausewitz and Contemporary War* (Oxford: Oxford University Press, 2007), 165.

47. Carl von Clausewitz, *On War*, trans. Colonel J. J. Graham (London: Routledge and Kegan Paul, 1968), 3:250. I am grateful to Olivia Garard for pointing me to this passage.

48. Ibid., 3:250–252.

49. Coker, *The Warrior Ethos*, 52.

50. Homer, *The Iliad*, 292.

51. Coker, *The Warrior Ethos*, 64; French, *The Code of the Warrior*, 40.

52. Frevert, *Men of Honour*, 2–3.

53. Chernow, *Alexander Hamilton*, 691.

54. For the account of Burr and Bentham, see ibid., 691, 94, 720, respectively.

55. Frevert, *Men of Honour*, 11.

56. Coker spends some time discussing war as a transformative experience for combatants. See Coker, *The Warrior Ethos*, 4–5.

57. Gray, *The Warriors*, 44.

58. I follow Clausewitz's terminology here. Though the character of war can change with technology, politics, and governance, the nature of war remains the same. There is, however, some interesting discussion on this question. See, for example, Christopher Mewett, "Understanding War's Enduring Nature Alongside Its Changing Character," *War on the Rocks*, 21 January 2014; Rosa Brooks, "Fighting Words," *Foreign Policy*, 4 February 2014; Michael C. Sirak, "ISR Revolution," *Air Force Magazine*, 1 June 2010, 37; Stephen L. McFarland, *A Concise History of the U.S. Air Force* (Washington DC: Air Force History and Museums Program, 1997), 40; Colin S. Gray, "War: Continuity in Change, and Change in Continuity," *Parameters* 40, no. 2 (2010): 90, 606.

59. Orville Wright, "Letter to C. M. Hitchcock," Flight's Future, Smithsonian Education. www.smithsonianeducation.org/educators/lesson_plans/wright/flights_future.html.

60. "Irresistible Force," *Popular Mechanics*, February 1992, 102–104.

61. For dynamite and the machine gun, see Kevin Kelly, *What Technology Wants* (New York: Viking, 2010), 191. For the atomic bomb, see Ruth Lewin Sime, *Lise Meitner: A Life in Physics*, California Studies in the History of Science (Berkeley: University of California Press, 1996), 13: 375.

62. Frevert, *Men of Honour*, 174.

63. Strawser, "Moral Predators,"356.

64. See, respectively, Suzy Killmister, "Remote Weaponry: The Ethical Implications," *Journal of Applied Philosophy* 25, no. 2 (2008): 122; Enemark, *Armed Drones and the Ethics of War*, 60; Chamayou, *Theory of the Drone*, 62.

65. Cantwell, "Beyond Butterflies," 82; Gusterson, *Drone*, 54; Chamayou, *Theory of the Drone*, 99; Grace E. Miller, "'Boom / [S]He Is Not': Drone Wars and the Vanishing Pilot," *War, Literature and the Arts* 29 (2017): 3–4.

66. These lyrics are reproduced with the permission of the authors, Robert "Trip" Raymond and Chris "Snooze" Kurek.

67. I use "GPS-guided" as shorthand. Joint direct attack munitions are guided by an internal navigation system and are aided by GPS.

68. Ryan Hill, "Have I Ever Been to War?," *Military Review*, January–February 2020. These stanzas are reprinted with the permission of *Military Review*, the Professional Journal of the US Army, Combined Arms Center, Fort Leavenworth, Kansas.

69. US Army, "Army Regulation 600-8-22: Military Awards" (2019), 110.

70. Tim Shea, "A Combat Badge Does Not a Soldier Make," in *The Angry Staff Officer* (blog), 1 February 2016. https://angrystaffofficer.com/2016/02/01/a-combat-badge-does-not-a-soldier-make.

71. Hanson, *Western Way of War*, 225–227.

72. Coker, *The Warrior Ethos*, 43.

73. Augustine, *City of God*, trans. Marcus Dods (New York: Random House, 1993), 683–684.

74. French, *Code of the Warrior*, 12.

## 3   THE MORALITY AND PSYCHOLOGY OF REMOTE WARFARE

1. This story was recounted to me by Captain Andrew, a pilot who had been in the same squadron at the time. This story is also mentioned in Blair and House, "Avengers in Wrath."

2. Philip Alston, "Report of the Special Rapporteur on Extrajudicial, Summary or Arbitrary Executions: Study on Targeted Killings," UN General Assembly, Human Rights Council, Agenda Item 3, 28 May 2010, 25, www2.ohchr.org/english/bodies/hrcouncil/docs/14session/A.HRC.14.24.Add6.pdf.

3. Mark Bowden, "The Killing Machines: How to Think About Drones," *Atlantic* 312, no. 2 (2013): 6–7.

4. Kiel Brennan-Marquez, "A Progressive Defense of Drones," *Salon*, 24 May 2013, www.salon.com/2013/05/24/a_progressive_defense_of_drones.

5. Laurie Calhoun, "The End of Military Virtue," *Peace Review* 23, no. 3 (2011): 382.

6. See, respectively, Royakkers and van Est, "The Cubicle Warrior"; Calhoun, "End of Military Virtue"; John Kaag, "Drones, Ethics and the Armchair Soldier," *Opinionator* (blog), *New York Times*, 17 March 2013.

7. Dave Grossman, *On Killing: The Psychological Cost of Learning to Kill in War and Society*, rev. ed. (New York: Little, Brown, 2009). Though Grossman has been an influential figure in military circles, his work has been criticized, especially for its reliance on S. L. A. Marshall's fieldwork, which has been controversial since it was first published in 1947. Samuel Lyman Atwood Marshall, *Men Against Fire: The Problem of Battle Command in Future War* (Washington, DC: Infantry Journal Press, 1947).

8. Grossman, *On Killing*, 98.

9. Gray, *The Warriors*, xviii.

10. Clausewitz, *On War*, 250.

11. See, for example, Mark Coeckelbergh, "Drones, Information Technology, and Distance: Mapping the Moral Epistemology of Remote Fighting," *Ethics and Information Technology* 15, no. 2 (2013): 91–92; Royakkers and van Est, "The Cubicle Warrior," 292; Jai Galliott, *Military Robots: Mapping the Moral Landscape*, Military and Defence Ethics (Farnham, UK: Ashgate, 2015), 136–138; P. Asaro, "Modeling the Moral User," *IEEE Technology and Society Magazine* 28, no. 1 (2009): 22; Gregory, "From a View to a Kill," 197–198.

12. Chamayou, *Theory of the Drone*, 110.

13. Bryant is often misidentified as a pilot. See, for example, Nicola Abé, "The Woes of an American Drone Operator," *Spiegel Online*, 14 December 2012; Brandon Bryant, "Former Nellis AFB Drone Operator on First Kill, PTSD, Being Shunned by Fellow Airmen," in *State of Nevada*, ed. Joe Schoenmann (Las Vegas: Nevada Public Radio, 2015); Lauren Walker, "Death from Above: Confessions of a Killer Drone Operator," *Newsweek*, 19 November 2015; Michelle Bentley, "Fetishised Data: Counterterrorism, Drone Warfare and Pilot Testimony," *Critical Studies on Terrorism* 11, no. 1 (2018): 92; Matthew Power, "Confessions of a Drone Warrior," *GQ*, 23 October 2013.

14. Power, "Confessions of a Drone Warrior."

15. Ibid.

16. Travis Norton, email correspondence with author, 27 June 2019.

17. Brandon Bryant et al., "Final Drone Letter," *Guardian*, 7 December 2015.

18. Christine Evans, "Drones, Projections, and Ghosts: Restaging Virtual War in *Grounded* and *You Are Dead. You Are Here*," *Theatre Journal* 67, no. 4 (2015): 667.

19. Kent David Drescher et al., "An Exploration of the Viability and Usefulness of the Construct of Moral Injury in War Veterans," *Traumatology* 17, no. 1 (2011): 8–9; Shira Maguen and Brett T. Litz, "Moral Injury in Veterans of War," *PTSD Research Quarterly* 23, no. 1 (2012): 2. Craig J. Bryan et al., "Measuring Moral Injury: Psychometric Properties of the Moral Injury Events Scale in Two Military Samples," *Assessment* 23, no. 5 (2016): 558.

20. Tine Molendijk, Eric-Hans Kramer, and Désirée Verweij, "Moral Aspects of 'Moral Injury': Analyzing Conceptualizations on the Role of Morality in Military Trauma," *Journal of Military Ethics* 17, no. 1 (2018): 2.

21. In a 2010 study of 2700 US Army veterans returning from combat, Shira Maguen and colleagues found that even after controlling for combat exposure, killing was a significant predictor of PTSD symptoms. Shira Maguen et al., "The Impact of Reported Direct and Indirect Killing on Mental Health Symptoms in Iraq War Veterans," *Journal of Traumatic Stress* 23, no. 1 (2010): 88. See also Wayne Chappelle et al., "Combat and Operational Risk Factors for Post-Traumatic Stress Disorder Symptom Criteria Among United States Air Force Remotely Piloted Aircraft 'Drone' Warfighters," *Journal of Anxiety Disorders* 62 (2019): 87; Irina Komarovskaya et al., "The Impact of Killing and Injuring Others on Mental Health Symptoms Among Police Officers," *Journal of Psychiatric Research* 45, no. 10 (2011); Rachel MacNair, *Perpetration-Induced Traumatic Stress: The Psychological Consequences of Killing*, Psychological Dimensions to War and Peace (London: Praeger, 2002).

22. See, especially, Wayne Chappelle et al., "Prevalence of High Emotional Distress and Symptoms of Post-Traumatic Stress Disorder in U.S. Air Force Active Duty Remotely Piloted Aircraft Operators: 2010 USAFAM Survey Results," final technical report, Wright Patterson Air Force Base, OH, 2012; Wayne Chappelle et al., "An Analysis of Post-Traumatic Stress Symptoms in United States Air Force Drone Operators," *Journal of Anxiety Disorders* 28, no. 5 (2014); Wayne Chappelle et al., "Emotional Reactions to Killing in Remotely Piloted Aircraft Crewmembers During and Following Weapon Strikes," *Military Behavioral Health* (2018); Chappelle et al., "Combat and Operational Risk Factors." The various studies do not represent a one-to-one comparison, primarily because from 2018 to 2019, Chappelle and his colleagues transitioned from the DSM-IV to the DSM-5. Because the PTSD symptom criteria changed from one manual to the next, the conclusions are not directly comparable. The 2019 study resulted in the highest percentage of crew members who met PTSD symptom criteria: 6.15 percent. The previous study yielded 4.3 percent under the DSM-IV.

23. The numerous studies of PTSD rates among returning veterans have relied on varied methodologies. The reported ranges of PTSD rates, therefore, can be quite different from one another. Various studies conducted by

Wayne Chappelle and his colleagues cite the following ranges: 7.7–8.7 percent; 14–16 percent; 10–18 percent; and a meta-analysis of numerous studies that resulted in a range of 4–17 percent. In their discussion of their own findings, however, Chappelle et al., "Combat and Operational Risk Factors," 87, 91, compare the 6.15 percent rate to the returning veteran range of 4–18 percent.

24. Michael D. Collins, "A Fear of Flying: Diagnosing Traumatic Neurosis Among British Aviators of the Great War," *First World War Studies* 6, no. 2 (2015): 191, 97.

25. H. Graeme Anderson, Oliver H. Gotch, and Lord Weir of Eastwood, *The Medical and Surgical Aspects of Aviation* (London: London and Norwich Press, 1919), vii.

26. Schultz, *Problem with Pilots*, 17.

27. See, for example, Collins, "Fear of Flying," 191; L. Broughton-Head, "Studies in the Medical Aspects of Aviation, with Some Observations upon the 'Flying Temperament'" (University of Glasgow, 1919), 30; Anderson, Gotch, and Lord Weir of Eastwood, *Medical and Surgical Aspects of Aviation*, 41.

28. Schultz, *Problem with Pilots*, 15.

29. Crudely summarized, these are (1) physical or mental exhaustion; (2) breakdown from a purely mental origin; (3) a toxic element that has been the deciding factor in the breakdown; (4) psychopathy; (5) disability from a purely physical cause; and (6) "malingerers."

30. Anderson, Gotch, and Lord Weir of Eastwood, *Medical and Surgical Aspects of Aviation*, 118.

31. Mark K. Wells, *Courage and Air Warfare: The Allied Aircrew Experience in the Second World War*, Cass Series—Studies in Air Power (London: Frank Cass, 1995), 60.

32. Miller, *Masters of the Air*, 132.

33. Collins, "Fear of Flying," 197. For additional references to "lacking moral fiber," see Miller, *Masters of the Air*, 126.

34. Miller, *Masters of the Air*, 127.

35. Wayne Chappelle et al., "Assessment of Occupational Burnout in United States Air Force Predator/Reaper 'Drone' Operators," *Military Psychology* 26, no. 5–6 (2014); Wayne Chappelle et al., "Assessment of Occupational Burnout in United States Air Force Predator/Reaper 'Drone' Operators," *Military Psychology* 26, no. 5–6 (2017); Chappelle et al., "Prevalence of High Emotional Distress."

36. Peter Lee, "The Distance Paradox: Reaper, the Human Dimension of Remote Warfare, and Future Challenges for the RAF," *Air Power Review* 21, no. 3 (2018): 113.

37. Joseph Campo, "From a Distance: The Psychology of Killing with Remotely Piloted Aircraft," PhD diss., School of Advanced Air and Space Studies, Maxwell AFB, AL, 2015, 8, 43–44. Emphasis in original.

38. Enemark, *Armed Drones and the Ethics of War*, 120.

39. Chamayou, *Theory of the Drone*, 117. Gusterson makes a similar point when he says this kind of "remote intimacy . . . plays havoc with Grossman's model." Gusterson, *Drone*, 72.

40. Gray, *The Warriors*, 234.

41. Miller, *Masters of the Air*, 2. For a similar account, see Holland, *Battle of Britain*, 238–239.

42. Jonathan Shay, *Achilles in Vietnam: Combat Trauma and the Undoing of Character*, trade pbk. ed. (New York: Scribner, 2003).

43. For the definition of *moral injury* I use here, see Maguen and Litz, "Moral Injury in Veterans of War," 1.

44. *Transgressive act* is Sheila Frankfurt's term. I use it because it is more succinct than the alternative, "acts that transgress deeply held beliefs." See Sheila Frankfurt, "An Empirical Investigation of Moral Injury in Combat Veterans," PhD diss., University of Minnesota, 2015, 1.

45. Military ethicist Pauline Shanks Kaurin argues along similar lines. She wants to maintain Shay's definition and suggests that some of the effects referred to as moral injury in recent work result instead from moral perfectionism, moral luck, or moral uncertainty. Pauline Shanks Kaurin, "Healing the Wounds of War: Moral Luck, Moral Uncertainty, and Moral Injury," *Strategy Bridge*, 5 January 2018.

46. "Combat is one of the very few experiences where trauma exposure comes not only through being the . . . victim of violence . . . but also through inflicting . . . violence and destruction upon others" (Drescher et al., "The Usefulness of the Construct of Moral Injury," 8). For other recent psychological studies on moral injury, see, for example, Frankfurt, "Empirical Investigation of Moral Injury"; Maguen et al., "The Impact of Reported Direct and Indirect Killing"; William P. Nash et al., "Psychometric Evaluation of the Moral Injury Events Scale," *Military Medicine* 178, no. 6 (2013); Shira Maguen et al., "The Impact of Killing in War on Mental Health Symptoms and Related Functioning," *Journal of Traumatic Stress* 22, no. 5 (2009).

47. Gray, *The Warriors*, 184.

48. Shira Maguen and Brett T. Litz, "Moral Injury in Veterans of War," 1; Sheila Frankfurt and Patricia Frazier, "A Review of Research on Moral Injury in Combat Veterans," *Military Psychology* 28, no. 5 (2016): 318.

49. For examples of studies in which veterans suffer moral injury for committing morally justified actions, see Maguen and Litz, "Moral Injury in Veterans of War," 86–90; Maguen et al., "The Impact of Killing in War," 435–443;

Drescher et al., "The Usefulness of the Construct of Moral Injury," 8–9. See also Nancy Sherman, *Afterwar: Healing the Moral Injuries of Our Soldiers* (New York: Oxford University Press, 2015).

50. Nadine Barclay, "Hunters Save Lives Through RPA Human Performance Team," US Air Force, news release, 30 June 2015, www.creech.af.mil /News/Article-Display/Article/669956/hunters-save-lives-through-rpa-human -performance-team/. See also Eyal Press, "The Wounds of the Drone Warrior," *New York Times Magazine*, 13 July 2018.

51. Chaplain (Major) Joel, interview with author, 20 March 2019, Creech Air Force Base, NV. Members of the Human Performance Team and some of the other service members I interviewed preferred to be cited by rank and first name only.

52. International law and US Joint Chiefs doctrine refer to chaplains as noncombatants. The *Department of Defense Law of War Manual* refers to chaplains as "a special category under the law of war." See, respectively, Geneva Convention, Additional Protocol I to the Geneva Conventions, 1977, (I) Article 24; Joint Chiefs of Staff, *Joint Guide 1-05: Religious Affairs in Joint Operations* (Washington, DC: Joint Staff, 2018), vii; US Department of Defense, *Department of Defense Law of War Manual* (Washington, DC: Office of the General Counsel, 2016), 128.

53. Master Sergeant Sean, interview with author, 14 March 2019, Shaw AFB, SC.

54. Lieutenant Colonel Richard, PhD, interview with author, 21 March 2019, Creech Air Force Base, NV.

55. Joseph L. Campo, "Distance in War: The Experience of MQ-1 and MQ-9 Aircrew," *Air and Space Power Journal* (2015): 7–8.

56. Peter Lee's study is an important outlier in that it did pay close attention to moral injury. Lee, "The Distance Paradox," 124–125.

57. Rebecca Grant, "The ROVER," *Air Force Magazine*, August 2013, 39–42.

58. Jason M. Brown, "Operating the Distributed Common Ground System: A Look at the Human Factor in Net-Centric Operations," *Air and Space Power Journal* 23, no. 4 (2009): 51–57.

59. Matthew Atkins, "Lt. Col. Matthew Atkins on 'The Personal Nature of War in High Definition,'" ed. Benjamin Wittes, *Lawfare*, 26 January 2014.

60. Sonia Kennebeck and Ines Hofmann Kanna, *National Bird*, Independent Lens documentary film, ed. Lois Vossen (Public Broadcasting System, 2017).

61. The study found that 74 percent of respondents experienced some kind of emotional response the first time they took a human life in a weapons

employment. Some 57 percent of the responses were positive; 32 percent were negative; and 22 percent were both positive and negative. Campo, "From a Distance," 121, 24, 32.

62. Ibid., 132.

63. Lieutenant Mister, interview with author, 21 March 2019, Creech AFB, NV.

64. For example, Gray says, "Such loyalty to the group is the essence of fighting morale. The commander who can preserve and strengthen it knows that all other psychological or physical factors are little in comparison." Gray, *The Warriors*, 40.

65. Squadron Leader Russ, interview with author, 22 March 2019, Creech AFB, NV.

66. Lieutenant Steven, interview with author, 22 March 2019, Creech AFB, NV.

67. Lieutenant Colonel Richard, PhD, interview with author, 21 March 2019, Creech Air Force Base, NV.

68. Major Maria, interview with author, 20 March 2019, Creech AFB, NV.

## 4   GOOD GUYS AND BAD GUYS

1. For Mehsud accounts, see Mayer, "The Predator War." For Aulaqi, see Jere Van Dyk, "Who Were the 4 U.S. Citizens Killed in Drone Strikes?," *CBS News*, 23 May 2013; Michael Epstein, "The Curious Case of Anwar Al-Aulaqi: Is Targeting a Terrorist for Execution by Drone Strike a Due Process Violation When the Terrorist Is a United States Citizen?," *Michigan State University College of Law Journal of International Law* 19 (2010). For Soleimani, see Michael Crowley, Falih Hassan, and Eric Schmitt, "U.S. Strike in Iraq Kills Qassim Suleimani, Commander of Iranian Forces," *New York Times*, 7 January 2020; Julian Borger and Martin Chulov, "US Kills Iran General Qassem Suleimani in Strike Ordered by Trump," *Guardian*, 3 January 2020.

2. The Obama administration defined imminent threats as "ongoing" threats. See Office of Legal Counsel, "Memorandum for the Attorney General Re: Applicability of Federal Criminal Laws and the Constitution to Contemplated Lethal Operations Against Shaykh Anwar Al-Aulaqi" (Washington, DC, 2010); Eric Holder, "Attorney General Eric Holder Speaks at Northwestern University School of Law" (Chicago: Office of Public Affairs, 2012).

3. This story came from Technical Sergeant Jesus, interview with author, 21 March 2019, Creech AFB, NV; and Julian, interview with author, 1 October 2020, Washington, DC.

4. Julian, interview with the author, 1 October 2020, Washington, DC.

5. Human Rights Watch, "Iraq: ISIS Escapees Describe Systematic Rape," 14 April 2015, www.hrw.org/news/2015/04/14/iraq-isis-escapees-describe -systematic-rape.

6. United Nations, "Charter of the United Nations," *Yale Law Journal* 55, no. 5 (1946): 1303.

7. US Army, Arlington National Cemetery, "Audie Murphy," www .arlingtoncemetery.mil/Explore/Notable-Graves/Medal-of-Honor-Recipients /World-War-II-MoH-recipients/Audie-Murphy.

8. Christopher J. Eberle, *Justice and the Just War Tradition: Human Worth, Moral Formation, and Armed Conflict* (New York: Routledge, 2016), 30; Sebastian Junger, *War* (New York: Twelve, 2010).

9. Though the terms *discrimination*, *proportionality*, and *necessity* are used in the laws of war and in international humanitarian law, the legal definitions differ from their definitions in just war thinking.

10. David S. Cloud, "Anatomy of an Afghan War Tragedy," *Los Angeles Times*, 10 April 2011.

11. Dan Lamothe, "Investigation: Friendly Fire Airstrike That Killed U.S. Special Forces Was Avoidable," *Washington Post*, 4 September 2014.

12. Matthew Rosenberg, "Pentagon Details Chain of Errors in Strike on Afghan Hospital," *New York Times*, 29 April 2016.

13. Chamayou employed an argument like this one in 2013. Renic made a similar but stronger argument in 2020. Chamayou, *Theory of the Drone*; Neil C. Renic, *Asymmetric Killing: Risk Avoidance, Just War, and the Warrior Ethos* (Oxford: Oxford University Press, 2020).

14. Sarah Pruitt, "Heroes of Pearl Harbor: George Welch and Kenneth Taylor," *History*, 28 November 2016; updated 7 December 2018, www.history.com /news/heroes-of-pearl-harbor-george-welch-and-kenneth-taylor.

15. One approach to just war theory, sometimes called *revisionist*, offers an account of how a combatant loses the right not to be killed; this view is different from the traditional view. But this distinction need not concern us here.

16. This argument comes from Renic, *Asymmetric Killing*, 10, 35, 39, 43, 47, 98, 122.

17. Judith Armatta, *Twilight of Impunity: The War Crimes Trial of Slobodan Milosevic* (Durham, NC: Duke University Press, 2010), 11.

18. On one account of just war theory, often called the revisionist view, a combatant loses the right not to be killed only if the person fights on the unjust side of a war. So, in this view, if the US war in Afghanistan in 2012 was morally justified, then the revisionist view is that Dakota is morally justified in harming the sniper, but the sniper is not morally justified in harming Dakota.

19. Churchill, "The Few."

20. See Campo, "From a Distance." See also Chapter 3.

21. US Congress, "Loss of Nationality by Native-Born or Naturalized Citizen; Voluntary Action; Burden of Proof; Presumptions," 8 US Code § 1481 (Ithaca, NY: Cornell Law School, 1952).

22. Dominic Tierney, "The Twenty Years' War," *Atlantic*, 23 August 2016.

## 5   HUMAN JUDGMENT AND REMOTE WARFARE

1. US Air Force, *Air Force Instruction 13-112: Joint Terminal Attack Controller (JTAC) Training* (Washington, CD: USAF, 2017), 5.

2. Though there is no way to be certain, there apparently were similar rules of engagement in Peter Lee's recounting of a British Reaper pilot and ground force in Afghanistan. Peter Lee, *Reaper Force: The Inside Story of Britain's Drone Wars* (London: John Blake Publishing, 2018), 280–282.

3. John A. Tirpak, "Bombers over Libya," *Air Force Magazine* 94 (2011): 37.

4. Over time, DoD guidance on which elements of the nine-line have to be read back have varied. In general, the read-back requirement can be found in Joint Chiefs of Staff, *Close Air Support,* Joint Publication 3-09.3 (Washington, DC: Joint Staff, 2014).

5. Lieutenant Briana, Skype interview with author, 1 May 2020.

6. Steven J. DeTeresa et al., *The Joint Improvised Explosive Device Defeat Organization: DoD's Fight Against IED' Today and Tomorrow*, US House of Representatives, Committee on Armed Services, Subcommittee on Oversight & Investigations, November 2008, 9, https://armedservices.house.gov/_cache /files/c/f/cfddccb2-fc15-4a3d-b7e3-50fe3ea68eca/D09F0BEF55D1B39D 2CC196408918781D.jieddo-report-11-08-vf.pdf.

7. Captain Briana, Skype interview with author, 1 May 2020.

8. Kean et al., *9/11 Commission Report*, 116–117.

9. Lawrence Wright, *The Looming Tower: Al-Qaeda's Road to 9/11* (London: Penguin, 2011), 321–322.

10. This was true of Tomahawk variants through the Block III, which required eighty hours of planning for a strike. The Block IV, which didn't reach initial operating capability until 2004, can be switched to a secondary target in flight but still requires an hour of mission planning for each launch. Defense Industry Daily Staff, "Tomahawk's Chops: XGM-109 Block Iv Cruise Missiles," *Defense Industry Daily*, 13 July 2020.

11. US Navy, "Tomahawk Cruise Missile," US Navy Office of Information, news release, 27 September 2021, https://www.navy.mil/Resources/Fact-Files/Display -FactFiles/Article/2169229/tomahawk-cruise-missile/; Shane, *Objective Troy*.

12. Nicholas Schmidle, "Getting Bin Laden," *New Yorker*, 1 August 2011; Steven Swinford, "Osama Bin Laden Dead: Blackout During Raid on Bin Laden Compound," *Telegraph* (London), 4 May 2011.

13. Swinford, "Osama Bin Laden Dead."

14. This quotation is from Swinford, "Osama Bin Laden Dead." The claim that killing bin Laden was a split-second decision is contested. For instance, Schmidle, "Getting Bin Laden," cites a special operations officer who said, "There was never any question of detaining or capturing him—it wasn't a split-second decision. No one wanted detainees." Because I am after the conceptual distinction between physical distance and the judgment gap, this disagreement can be set aside.

15. For "robotic," see David Kilcullen testimony in US House of Representatives, Committee on Armed Services, *Effective Counterinsurgency*; Hastings, "Gatwick Drone Shambles"; Petrenko, "Between Berserksgang and the Autonomous Weapons Systems," 82; Royakkers and van Est, "The Cubicle Warrior," 289; Riza, "Two-Dimensional Warfare," 259; Rae and Crist, *Analyzing the Drone Debates*, 91; Sparrow, "War Without Virtue?," 86; Kaag and Kreps, *Drone Warfare*. For "semi-autonomous," see Brunstetter and Braun, "The Implications of Drones," 338; Kaag and Kreps, *Drone Warfare*, vii; Royakkers and van Est, "The Cubicle Warrior," 293.

16. Lieutenant Clifton, interview with author, 14 March 2019, Shaw AFB, SC.

17. Ibid.

18. Technical Sergeant Megan, interview with author, 14 March 2019, Shaw AFB, SC.

19. Master Sergeant Sean, interview with author, 14 March 2019, Shaw AFB, SC.

20. Kennebeck and Kanna, *National Bird*.

21. Technical Sergeant Megan, interview with author, 14 March 2019, Shaw AFB, SC.

22. US Air Force, *Annex 3-60: Targeting*.

23. Captain Shaun, interview with author, 15 March 2019, Shaw AFB, SC.

24. Cheryl J. Roby and David S. Alberts, *NATO NEC C2 Maturity Model* (Washington, DC: Center for Advanced Concepts and Technology [ACT], 2010), xvi; Curtis M. Scaparrotti and Denis Mercier, *Framework for Future Alliance Operations*, Keeping the Edge, North Atlantic Treaty Organization, 2018, www.act.nato.int/images/stories/media/doclibrary/180514_ffao18.pdf; UK Ministry of Defence, *Future of Command and Control* (Swindon, UK: Development, Concepts and Doctrine Centre, 2017), 6, 18, 37; Jim Storr, "A Command Philosophy for the Information Age: The Continuing Relevance of Mission Command," *Defence Studies* 3, no. 3 (2003).

## 6   IT'S "HARD WORK TO BE EXCELLENT"

1. Cara Daggett, "Drone Disorientations," *International Feminist Journal of Politics* 17, no. 3 (2015): 369; Michael Byrnes, "Dark Horizon: Airpower

Revolution on a Razor's Edge—Part Two of the 'Nightfall' Series," *Air & Space Power Journal* 29, no. 5 (2015): 51n. See also Marina Powers, "Sticks and Stones: The Relationship Between Drone Strikes and Al-Qaeda's Portrayal of the United States," *Critical Studies on Terrorism* 7, no. 3 (2014); Kenneth Anderson, "Predators over Pakistan," *Weekly Standard* 15, no. 24 (2010): 26–28. Anderson and Byrnes identify this charge as a trend in the literature without adopting the view themselves.

2. Plato associates courage with "fighting in armor" in the *Laches*. Aristotle says that "someone is called fully brave if he is intrepid in facing a fine death. . . . And this is above all true of the dangers of war." Plato, "Laches," in *The Collected Dialogues of Plato*, ed. Edith Hamilton, Bollingen Series (Princeton, NJ: Princeton University Press, 2002), 71:134. Aristotle, *Nicomachean Ethics*, trans. Terence Irwin (Indianapolis: Hackett Publishing Company, 2000), 41.

3. Technical Sergeant Megan, interview with author, 14 March 2019, Shaw AFB, SC.

4. Daggett, "Drone Disorientations," 369; Byrnes, "Dark Horizon," 51n. See also Powers, "Sticks and Stones."

5. Anderson, "Predators over Pakistan," 26–28. As I read it, Anderson identifies this charge as a trend in the literature without adopting the view himself.

6. Rowan Scarborough, "Pentagon Uproar over Panetta's Hero Medals for Drone Operators, Cybersleuths," *Washington Times*, 15 February 2013.

7. Riza, *Killing Without Heart*.

8. Sparrow, "War Without Virtue?," 81–105. See also Jesse Kirkpatrick, "Reply to Sparrow: Martial Courage—or Merely Courage?," *Journal of Military Ethics* 14, no. 3–4 (2015); Robert Sparrow, "Martial and Moral Courage in Tele-operated Warfare: A Commentary on Kirkpatrick," *Journal of Military Ethics* 14, no. 3–4 (2015): 220–227.

9. See Peter Olsthoorn, *Military Ethics and Virtues: An Interdisciplinary Approach for the 21st Century* (London and New York: Routledge, 2011), 6.

10. Philippa Foot, *Virtues and Vices and Other Essays in Moral Philosophy* (Berkeley: University of California Press, 1978), 1; Rosalind Hursthouse, *On Virtue Ethics* (Oxford: Oxford University Press, 1999), 1. The primary virtue ethics text from the ancient Greek period, and the one that remains the primary source for Western virtue ethics, is a collection of Aristotle's lecture notes called *The Nicomachean Ethics*. See Aristotle, *Nicomachean Ethics*, xiv–xv.

11. Hursthouse, *On Virtue Ethics*, 2, 7.

12. Ralph McInerny and John O'Callaghan, "Saint Thomas Aquinas," in *The Stanford Encyclopedia of Philosophy*, ed. Edward N. Zalta (Stanford, CA: Stanford University, 2018).

13. For the claim that virtue ethics is particularly important to military contexts, see Hilliard Aronovitch, "Good Soldiers, a Traditional Approach," *Journal of Applied Philosophy* 18, no. 1 (2001); René Moelker and Peter Olsthoorn, "Virtue Ethics and Military Ethics," *Journal of Military Ethics* 6, no. 4 (2007); Peter Olsthoorn, "Courage in the Military: Physical and Moral," *Journal of Military Ethics* 6, no. 4 (13 December 2007): 270–279; Olsthoorn, *Military Ethics and Virtues*; Paul Robinson, "Magnanimity and Integrity as Military Virtues," *Journal of Military Ethics* 6, no. 4 (2007); Jessica Wolfendale, "What Is the Point of Teaching Ethics in the Military?," in *Ethics Education in the Military*, ed. Paul Robinson, Nigel de Lee, and Don Carrick (New York: Ashgate, 2008).

14. Aristotle, *Nicomachean Ethics*, 9–16; Hursthouse, *On Virtue Ethics*, 9; Robert Merrihew Adams, *A Theory of Virtue: Excellence in Being for the Good* (Oxford: Clarendon Press, 2006), 49.

15. Aristotle, *Nicomachean Ethics*, 19.

16. Ibid., 20.

17. See, for example, Robin de Peyer, "Sherlock Star Benedict Cumberbatch Saves Deliveroo Cyclist Being Violently Mugged by Gang of Thugs," *Evening Standard*, 2 June 2018; Telegraph Reporters, "Benedict Cumberbatch Fights Off Four Muggers Who Attacked Deliveroo Cyclist near Baker Street," *Telegraph* (London), 1 June 2018; Jack Shepherd, "Benedict Cumberbatch Hailed a 'Hero' After Fending Off Four Muggers Attacking a Deliveroo Cyclist," *Independent* (London), 2 June 2018.

18. The preceding quotes are taken from de Peyer, "Sherlock Star Benedict Cumberbatch."

19. This view is from Sparrow, "Martial and Moral Courage," 224.

20. Kaurin, *Warrior, Military Ethics*, 16.

21. Chamayou likewise argues that remote warfare has redefined martial virtue. Chamayou, *Theory of the Drone*, 101.

22. Major Jordan, interview with author, 19 December 2016, Randolph AFB, TX.

23. Kirkpatrick, "Drones and the Martial Virtue Courage," 205–206.

24. Gray, *The Warriors*, 40.

25. Blair and House, "Avengers in Wrath."

26. Insinna, "Predators and Reapers Prove Close-Air Support Bona-Fides."

27. There is some considerable irony in that Michael Corleone was also, in fact, a US Marine.

28. Technical Sergeant Megan, interview with author, 14 March 2019, Shaw AFB, SC.

29. Staff Sergeant Jackson, interview with author, 20 March 2019, Creech AFB, NV.

30. Richard D. Winters and Cole C. Kingseed, *Beyond Band of Brothers: The War Memoirs of Major Dick Winters* (New York: Penguin, 2008), 291.

31. Marlantes, *What It Is Like to Go to War*, 40–41.

32. This and subsequent references to Luck are taken from Hans von Luck, "The End in North Africa," in *Experience of War: An Anthology of Articles from MHQ, the Quarterly Journal of Military History*, ed. Robert Cowley (New York: W. W. Norton and Company, 1992), 430–442.

33. Ibid., 437.

34. For the quoted text, see Holland, *The Battle of Britain*, 305. On the same page, Holland explains that before the Hawk engagement, Hans had "three aerial victories to his name." For Hans's sixty confirmed air-to-air victories, see David T. Zabecki, ed., *Germany at War: 400 Years of Military History* (Santa Barbara, CA: ABC-CLIO, LLC, 2014), 4:1617.

35. Or at least reduced risk of attack from Allied aircraft. Hans undoubtedly accepted additional risk of repercussions from the Luftwaffe.

36. Aristotle, *Nicomachean Ethics*, 29.

## 7   WHAT COMES NEXT?

1. See, for example, Micah Zenko and Sarah Kreps, *Limiting Armed Drone Proliferation. Council on Foreign Relations* (New York: Council on Foreign Relations, 2014); Ulrike Esther Franke, "The Global Diffusion of Unmanned Aerial Vehicles (UAVs), or 'Drones,'" in *Precision Strike Warfare and International Intervention: Strategic, Ethico-Legal, and Decisional Implications*, ed. Mike Araronson, Wali Aslam, Tom Dyson, and Regina Rauxloh, 52–72 (London: Routledge, 2015); Dan Gettinger, Arthur Holland Michel, Alex Pasternack, Jason Koebler, Shawn Musgrave, and Jared Rankin, *The Drone Primer: A Compendium of the Key Issues* (New York: Annandale-On-Hudson, 2014), https://dronecenter.bard.edu/the-drone-primer-announcement; Rebecca J. Johnson, "The Wizard of Oz Goes to War: The Unmanned Systems in Counterinsurgency," in *Killing by Remote Control: The Ethics of an Unmanned Military*, ed. Bradley Jay Strawser, 154–178 (Oxford: Oxford University Press, 2013).

2. See, for example, Ben Hubbard, Pako Karasz, and Stanley Reed, "Two Major Saudi Oil Installations Hit by Drone Strike, and U.S. Blames Iran," *New York Times*, 14 September 2019. Some claim the attack was conducted by both remotely piloted aircraft and cruise missiles. Erin Cunningham, "Iran's Gamble: Analysts Say Brazen Attack Aimed to Pressure U.S. with Little Fear of Reprisal," *Washington Post*, 20 September 2019.

3. Shane Harris, Erin Cunningham, and Kareem Fahim, "Trump Stops Short of Directly Blaming Iran for Attack on Saudi Oil Facilities," *Washington Post*, 17 September 2019.

4. Catalano Ewers et al., *Drone Proliferation* 2, 12. See also Bergen et al., *World of Drones*; Natasha Turak, "Pentagon Is Scrambling as China 'Sells the Hell out of' Armed Drones to US Allies," *CNBC*, 21 February 2019.

5. *Area of active hostilities* was originally an Obama administration term, though its use persisted under the Trump administration. Critics argue that because the term is not recognized in international law, it is too nebulous to serve as a meaningful constraint on the use of force. See Brianna Rosen, "To End the Forever Wars, Rein in the Drones," *Just Security*, 16 February 2021; Charlie Savage, "Trump Revokes Obama-Era Rule on Disclosing Civilian Casualties from U.S. Airstrikes Outside War Zones," *New York Times*, 6 March 2019.

6. US Congress, Public Law 107-40 (Joint Resolution to Authorize the Use of United States Armed Forces Against Those Responsible for the Recent Attacks Launched Against the United States), 107th Congress, 2001.

7. US Congress, Public Law 112-81 (National Defense Authorization Act for Fiscal Year 2012), 112th Congress, 2012.

8. Congressional approval for the US invasion of Iraq came through a second resolution, the 2002 AUMF.

9. Crowley, Hassan, and Schmitt, "U.S. Strike in Iraq."

10. According to one poll, 87 percent thought killing bin Laden during the raid was justified. Robert Burns and Jennifer Agiesta, "AP-GfK Poll: Bin Laden Killing Was Justified," *San Diego Union-Tribune*, 11 May 2011.

11. Kenneth R. Himes, *Drones and the Ethics of Targeted Killing* (Lanham, MD: Rowman & Littlefield, 2016), 3–6.

12. Philip Bump, "Why the Administration Claims That Soleimani Killed Hundreds of Americans," *Washington Post*, 8 January 2020.

13. Yuhas, "Airstrike That Killed Suleimani"; Schwartz, "Who Was the Iraqi Commander?"

14. The DoD's AI strategy says that AI is "the ability of machines to perform tasks that normally require human intelligence." US Department of Defense, "Artificial Intelligence Strategy," 5.

15. I choose this example because the US Department of Veterans Affairs has one of the largest data repositories in US health care and is actively pursuing data management and advanced data analytics to predict patient health-care needs. Stephan D. Fihn et al., "Insights from Advanced Analytics at the Veterans Health Administration," *Health Affairs* 33, no. 7 (2014): 1203–1211.

16. Ibid., 1206.

17. This kind of work is already underway in health-care research. See, for example, Thomas Davenport and Ravi Kalakota, "The Potential for Artificial Intelligence in Healthcare," *Future Healthcare Journal* 6, no. 2 (2019); Fei Jiang

et al., "Artificial Intelligence in Healthcare: Past, Present and Future," *Stroke and Vascular Neurology* 2, no. 4 (2017).

18. Cheryl Pellerin, "Project Maven to Deploy Computer Algorithms to War Zone by Year's End," US Department of Defense, news release, 21 July 2017, www.defense.gov/Explore/News/Article/Article/1254719/project-maven-to-deploy-computer-algorithms-to-war-zone-by-years-end.

19. Scott Shane and Daisuke Wakabayashi, "'The Business of War': Google Employees Protest Work for the Pentagon," *New York Times*, 4 April 2018; Drew Harwell, "Google to Drop Pentagon AI Contract After Employee Objections to the 'Business of War,'" *Washington Post*, 1 June 2018; Pellerin, "Project Maven to Deploy Computer Algorithms"; Douglas MacMillan, "Google Won't Seek to Renew Pentagon Contract after Internal Backlash," *Wall Street Journal*, 1 June 2018. More recently, Google has signaled that it might resume working with the DoD. Daisuke Wakabayashi and Kate Conger, "Google Wants to Work with the Pentagon Again, Despite Employee Concerns," *New York Times*, 3 November 2021.

20. Robert Work, "Establishment of an Algorithmic Warfare Cross-Functional Team (Project Maven)," memorandum of the Deputy Secretary of Defense for Secretaries of the Military Departments et al., Washington, DC, 26 April 2017; Pellerin, "Project Maven."

21. The exact numbers are complicated because for years the US Air Force measured the workforce needed to support Reaper operations in twenty-four-hour blocks called *combat air patrols*. To sustain twenty-four-hour coverage over a target area, the air force needs a certain number of pilots, sensor operators, aircraft, and intelligence analysts. According to one RAND study, a single twenty-four-hour sortie requires eighty-four intelligence analysts. There have been, at most, sixty-five sorties airborne at once. According to the study, more than five thousand personnel would have been required just to support the processing, exploitation, and dissemination of Predator and Reaper video. Lance Menthe et al., *The Future of Air Force Motion Imagery Exploitation: Lessons from the Commercial World* (Santa Monica, CA: RAND 2012).

22. Israel Aerospace Industries, "Harpy: Autonomous Weapon for All Weather"; Dan Gettinger and Arthur Holland Michel, "Loitering Munitions," Center for the Study of the Drone, Bard College, 2017, https://dronecenter.bard.edu/files/2017/02/CSD-Loitering-Munitions.pdf.

23. Missile Defense Project, "Brimstone," Center for Strategic and International Studies Missile Defense Project, 6 December 2017; updated 30 July 2021. https://missilethreat.csis.org/missile/brimstone.

24. BAE Systems, "Taranis," BAE Systems, information page, n.d. www.baesystems.com/en/product/taranis; Jürgen Altmann and Frank Sauer,

"Autonomous Weapon Systems and Strategic Stability," *Survival* 59, no. 5 (2017).

25. US Air Force Research Lab (AFRL), "Skyborg: Open . . . Resilient . . . Autonomous," AFRL, 2020. https://afresearchlab.com/technology/vanguards /successstories/skyborg.

26. Kyle Mizokami, "Air Force Tests New 'Loyal Wingman' Sidekick Drone for Combat," *Popular Mechanics*, 7 March 2019.

27. Defense Advanced Research Projects Agency (DARPA), "AlphaDog-fight Trials Foreshadow Future of Human-Machine Symbiosis," DARPA, news release, 26 August 2020, www.darpa.mil/news-events/2020-08-26.

28. Ashton B. Carter, "Autonomy in Weapon Systems," US Department of Defense Directive 3000.09, Washington, DC, 21 November 2012, www.esd .whs.mil/Portals/54/Documents/DD/issuances/dodd/300009p.pdf.

29. US Department of Defense (DoD), "DoD Adopts Ethical Principles for Artificial Intelligence," DoD, news release, 24 February 2020, www .defense.gov/Newsroom/Releases/Release/Article/2091996/dod-adopts-ethical -principles-for-artificial-intelligence.

30. Mattis, *Summary of the National Defense Strategy*.

31. Charles Q. Brown Jr., "Accelerate Change or Lose," US Air Force, Washington, DC, August 2020, 3, www.airforcemag.com/app/uploads/2020/09 /CSAF-22-Strategic-Approach-Accelerate-Change-or-Lose-31-Aug-2020.pdf.

32. Dana W. White and Kenneth F. McKenzie Jr., "Department of Defense Press Briefing by Pentagon Chief Spokesperson Dana W. White and Joint Staff Director Lt. Gen. Kenneth F. McKenzie Jr. in the Pentagon Briefing Room," DoD, news release, 14 April 2018, www.defense.gov/Newsroom/Transcripts /Transcript/Article/1493749/department-of-defense-press-briefing-by -pentagon-chief-spokesperson-dana-w-whit.

33. For reporting that the extended-range variant was used in the Syria strikes, see John A. Tirpak and Brian Everstine, "Syria Strike Marks Combat Debut for JASSM-ER," *Air Force Magazine*, 15 April 2018. For estimated ranges, see Missile Defense Project, "JASSM / JASSM-ER (AGM-158A/B)," Center for Strategic and International Studies, 6 October 2016; updated 30 July 2021, https://missilethreat.csis.org/missile/jassm.

34. US Space Force, *Spacepower: Doctrine for Space Forces*, Space Capstone Publication (Washington, DC: US Space Force, 2020), 31, 32.

# REFERENCES

Abé, Nicola. "The Woes of an American Drone Operator." *Spiegel Online*, 14 December 2012.

Adams, Robert Merrihew. *A Theory of Virtue: Excellence in Being for the Good.* Oxford: Clarendon Press, 2006.

Alston, Philip. "Report of the Special Rapporteur on Extrajudicial, Summary or Arbitrary Executions: Study on Targeted Killings." United Nations General Assembly, Human Rights Council, Agenda Item 3. 28 May 2010. www2.ohchr.org/english/bodies/hrcouncil/docs/14session/A.HRC.14.24 .Add6.pdf.

Altmann, Jürgen, and Frank Sauer. "Autonomous Weapon Systems and Strategic Stability." *Survival* 59, no. 5 (2017): 117–142.

Ambrose, Stephen E. *Band of Brothers.* London: Simon and Schuster UK, 2017.

Anderson, H. Graeme, Oliver H. Gotch, and Lord Weir of Eastwood. *The Medical and Surgical Aspects of Aviation.* London: London and Norwich Press, 1919.

Anderson, Kenneth. "Predators over Pakistan." *Weekly Standard* 15, no. 24 (8 March 2010): 26–34.

Aristotle. *Nicomachean Ethics.* Translated by Terence Irwin. Indianapolis: Hackett Publishing Company, 2000.

Armatta, Judith. *Twilight of Impunity: The War Crimes Trial of Slobodan Milosevic.* Durham, NC: Duke University Press, 2010.

Arnold, H. H., and Ira Eaker. *This Flying Game.* New York: Funk & Wagnalls, 1936.

Aronovitch, Hilliard. "Good Soldiers, a Traditional Approach." *Journal of Applied Philosophy* 18, no. 1 (2001): 13–23.

Asaro, P. "Modeling the Moral User." *IEEE Technology and Society Magazine* 28, no. 1 (2009): 20–24.

Asch, Alfred, Hugh R. Graff, and Thomas A. Ramey. *The Story of the Four Hundred and Fifty-Fifth Bombardment Group (H) WWII: Flight of the Vulgar Vultures.* Appleton, WI: Graphic Communications Center, 1991.

Atkins, Matthew. "Lt. Col. Matthew Atkins on 'The Personal Nature of War in High Definition.'" Edited by Benjamin Wittes. *Lawfare*, 26 January 2014.

Augustine. *City of God.* Translated by Marcus Dods. New York: Random House, 1993.

BAE Systems. "Taranis." BAE Systems, information page, n.d. www.bae systems.com/en/product/taranis.

Baldwin, Hanson W. "The 'Drone': Portent of Push-Button War; Recent Operations Point to the Pilotless Plane as a Formidable Weapon in War's New Armory." *New York Times*, 25 August 1946.

Barber, James. "Jake Gyllenhaal Set to Play Medal of Honor Recipient John Chapman in 'Combat Control.'" *Military.com*, 22 March 2021.

Barclay, Nadine. "Hunters Save Lives Through RPA Human Performance Team." US Air Force, news release, 30 June 2015. www.creech.af.mil /News/Article-Display/Article/669956/hunters-save-lives-through-rpa -human-performance-team.

Benjamin, Medea. *Drone Warfare: Killing by Remote Control*, updated ed. London: Verso, 2013.

Bentley, Michelle. "Fetishised Data: Counterterrorism, Drone Warfare and Pilot Testimony." *Critical Studies on Terrorism* 11, no. 1 (2018): 88–110.

Bergen, Peter, David Sterman, Melissa Salyk-Virk, Christopher Mellon, Alyssa Sims, and Albert Ford. *World of Drones: Examining the Proliferation, Development, and Use of Armed Drones.* Washington, DC: New America, 2017; updated 2019. www.newamerica.org/in-depth/world-of-drones.

Berger, David H. "Commandant's Planning Guidance: 38th Commandant of the Marine Corps." Washington, DC, 2019. www.marines.mil/Portals/1 /Publications/Commandant's%20Planning%20Guidance_2019.pdf.

Blair, Dave, and Karen House. "Avengers in Wrath: Moral Agency and Trauma Prevention for Remote Warriors." *Lawfare*, 12 November 2017.

Borger, Julian, and Martin Chulov. "US Kills Iran General Qassem Suleimani in Strike Ordered by Trump." *Guardian*, 3 January 2020.

Bowden, Mark. "The Killing Machines: How to Think About Drones." *Atlantic*, 18 August 2013, 6–7.

Boyle, Michael J. "The Costs and Consequences of Drone Warfare." *International Affairs* 89, no. 1 (2013): 1–29.

Boyne, Walter J. *Beyond the Wild Blue: A History of the US Air Force, 1947–2007.* New York: Macmillan, 2007.

———. "Creech." *Air Force Magazine*, 1 March 2005.

Brennan, Frank. "The Lonely Impulse of Delight." *Journal of Palliative Care* 28, no. 4 (2012): 297–300.

Brennan-Marquez, Kiel. "A Progressive Defense of Drones." *Salon*, 24 May 2013. www.salon.com/2013/05/24/a_progressive_defense_of_drones.

Brooks, Rosa. "Fighting Words." *Foreign Policy*, 4 February 2014.

———. *How Everything Became War and the Military Became Everything: Tales from the Pentagon*. New York: Simon & Schuster, 2016.

Broshear, Nathan. "Airmen Fly Predator in Controlled Airspace over Haiti." US Air Force, 12th Air Force Southern Public Affairs, 29 January 2010. www.af.mil/News/Article-Display/Article/117786/airmen-fly-predator-in-controlled-airspace-over-haiti.

Broughton-Head, L. "Studies in the Medical Aspects of Aviation, with Some Observations upon the 'Flying Temperament.'" Glasgow: University of Glasgow, 1919.

Brown, Charles Q., Jr. "Accelerate Change or Lose." US Air Force, Washington, DC, August 2020, www.airforcemag.com/app/uploads/2020/09/CSAF-22-Strategic-Approach-Accelerate-Change-or-Lose-31-Aug-2020.pdf.

Brown, Jason M. "Operating the Distributed Common Ground System: A Look at the Human Factor in Net-Centric Operations." *Air and Space Power Journal* 23, no. 4 (Winter 2009): 51–57.

Brunstetter, Daniel, and Megan Braun. "The Implications of Drones on the Just War Tradition." *Ethics & International Affairs* 25, no. 03 (2011): 337–358.

Bryan, Craig J., Annabelle O. Bryan, Michael D. Anestis, Joye C. Anestis, Bradley A. Green, Neysa Etienne, Chad E. Morrow, and Bobbie Ray-Sannerud. "Measuring Moral Injury: Psychometric Properties of the Moral Injury Events Scale in Two Military Samples." *Assessment* 23, no. 5 (2016): 557–570.

Bryant, Brandon. "Former Nellis AFB Drone Operator on First Kill, PTSD, Being Shunned by Fellow Airmen." In *State of Nevada*, edited by Joe Schoenmann. Las Vegas: Nevada Public Radio, 2015.

Bryant, Brandon, Cian Westmoreland, Stephen Lewis, and Michael Haas. "Final Drone Letter." *Guardian*, 7 December 2015.

Budiansky, Stephen. *Oliver Wendell Holmes: A Life in War, Law, and Ideas*. New York: W. W. Norton & Company, 2019.

Builder, Carl H. *The Army in the Strategic Planning Process: Who Shall Bell the Cat?* Santa Monica, CA: RAND, 1987. www.rand.org/pubs/reports/R3513.html#download.

Bump, Philip. "Why the Administration Claims That Soleimani Killed Hundreds of Americans." *Washington Post*, 8 January 2020.

Bundesstelle für Flugunfalluntersuchung (German Federal Bureau of Aircraft Accident Investigation). *Investigation Report: BFU EX010-11.* Braunschweig, Germany: Bundesstelle für Flugunfalluntersuchung, 2011. www.bfu-web .de/EN/Publications/Investigation%20Report/2011/Report_11_EX010 _B777_Munic.pdf?__blob=publicationFile.

Burns, Robert, and Jennifer Agiesta. "AP-GfK Poll: Bin Laden Killing Was Justified." *San Diego Union-Tribune,* 11 May 2011.

Bush, George W. "Selected Speeches of President George W. Bush." National Archives and Records Administration, Washington, DC, 2008.

Byrnes, Michael. "Dark Horizon: Airpower Revolution on a Razor's Edge— Part Two of the 'Nightfall' Series." *Air & Space Power Journal* 29, no. 5 (2015): 31–56.

Calhoun, Laurie. "Drone Killing and the Disastrous Doctrine of Double Effect." *Peace Review* 27, no. 4 (2015): 440–447.

———. "The End of Military Virtue." *Peace Review* 23, no. 3 (2011): 377–386.

Campo, Joseph. "From a Distance: The Psychology of Killing with Remotely Piloted Aircraft." PhD diss., School of Advanced Air and Space Studies, Maxwell Air Force Base, AL, 2015.

———. "Distance in War: The Experience of MQ-1 and MQ-9 Aircrew." *Air & Space Power Journal* (2015).

Cantwell, Houston R. "Beyond Butterflies: MQ-1 Predator and the Evolution of Unmanned Aerial Vehicles in Air Force Culture," Thesis, School of Advanced Air and Space Studies, 2007.

Carter, Ashton B. "Autonomy in Weapon Systems." US Department of Defense directive. Washington, DC, 21 November 2012. www.esd.whs.mil /Portals/54/Documents/DD/issuances/dodd/300009p.pdf.

Catalano Ewers, Elisa, Lauren Fish, Michael C. Horowitz, Alexandra Sander, and Paul Scharre. *Drone Proliferation: Policy Choices for the Trump Administration.* Center for a New American Security, Washington, DC, 1 June 2017.

Chamayou, Grégoire. *A Theory of the Drone.* Translated by Janet LLoyd. New York: New Press, 2013.

Chapa, Joseph O. "Remotely Piloted Aircraft and War in the Public Relations Domain." *Air & Space Power Journal* 28, no. 5 (September–October 2014): 29–46.

Chappelle, Wayne, Tanya Goodman, Laura Reardon, and Lillian Prince. "Combat and Operational Risk Factors for Post-Traumatic Stress Disorder Symptom Criteria Among United States Air Force Remotely Piloted Aircraft 'Drone' Warfighters." *Journal of Anxiety Disorders* 62 (11 January 2019): 86–93.

Chappelle, Wayne, Tanya Goodman, Laura Reardon, and William Thompson. "An Analysis of Post-Traumatic Stress Symptoms in United States Air Force Drone Operators." *Journal of Anxiety Disorders* 28, no. 5 (June 2014): 480–487.

Chappelle, Wayne, Kent McDonald, Lillian Prince, Tanya Goodman, Bobbie N. Ray-Sannerud, and William Thompson. "Assessment of Occupational Burnout in United States Air Force Predator/Reaper 'Drone' Operators." *Military Psychology* 26, no. 5–6 (2014): 376–385.

Chappelle, Wayne, Julie Swearengen, Tanya Goodman, Lillian Prince, and William Thompson. "Reassessment of Occupational Health Among US Air Force Remotely Piloted Aircraft (Drone) Operators." Report, US Air Force Research Laboratory, April 2017.

Chappelle, Wayne, Kent McDonald, William Thompson, and Julie Swearengen. "Prevalence of High Emotional Distress and Symptoms of Post-Traumatic Stress Disorder in U.S. Air Force Active Duty Remotely Piloted Aircraft Operators: 2010 USAFAM Survey Results." Wright Patterson Air Force Base, OH, 2012.

Chappelle, Wayne, Emily Skinner, Tanya Goodman, Julie Swearengen, and Lillian Prince. "Emotional Reactions to Killing in Remotely Piloted Aircraft Crewmembers During and Following Weapon Strikes." *Military Behavioral Health* (2018): 1–11.

Cheney, Richard B., and Liz Cheney. *In My Time: A Personal and Political Memoir.* New York, London: Threshold Editions, 2011.

Chernow, Ron. *Alexander Hamilton.* New York: Penguin Press, 2004.

Chivers, C. J. "A Changed Way of War in Afghanistan's Skies." *New York Times,* 15 January 2012.

———. *The Fighters: Americans in Combat in Afghanistan and Iraq.* New York: Simon and Schuster, 2018.

Churchill, Winston. "The Few: Churchill's Speech to the House of Commons, August 20, 1940." Churchill Society, London. www.churchill-society -london.org.uk/thefew.html.

Clark, Colin. "Reaper Drones: The New Close Air Support Weapon." *Breaking Defense,* 10 May 2018.

Clausen, Christian. "Combat RPAs Integral in Defeating ISIS." US Air Force, 432nd Wing/432nd Air Expeditionary Wing Public Affairs, 5 December 2017, www.af.mil/News/Article-Display/Article/1388206 /combat-rpas-integral-in-defeating-isis.

———. "Next Level of RPA Operations, USAFCENT Commander Recognizes Airmen." US Air Force, 432nd Wing/432nd Air Expeditionary Wing Public Affairs, 10 January 2018, www.creech.af.mil/News/Article

-Display/Article/1412436/next-level-of-rpa-operations-usafcent-commander
-recognizes-airmen.

Clausewitz, Carl von. *On War*. Translated by Michael Howard and Peter Paret. Oxford: Oxford University Press, 2008.

———. *On War*. Translated by Colonel J. J. Graham. 3 vols. London: Routledge and Kegan Paul, 1968.

Cloud, David S. "Anatomy of an Afghan War Tragedy." *Los Angeles Times*, 10
April 2011.

Clutton-Brock, Arthur, and André Chevrillon. *Letters from a Soldier of France
1914–1915: Wartime Letters from France*. South Yorkshire, UK: Pen and
Sword, 2014.

Coady, C. A. J. *Morality and Political Violence*. Cambridge: Cambridge University Press, 2008.

Cockburn, Andrew. *Kill Chain: The Rise of the High-Tech Assassins*. London:
Verso, 2015.

Coeckelbergh, Mark. "Drones, Information Technology, and Distance: Mapping the Moral Epistemology of Remote Fighting." *Ethics and Information
Technology* 15, no. 2 (2013): 87–98.

Coker, Christopher. "Ethics of Counterinsurgency." In *The Routledge Handbook of Insurgency and Counterinsurgency*, edited by Paul B. Rich and Isabelle Duyvesteyn, 119–127. London: Routledge, 2012.

———. *The Warrior Ethos: Military Culture and the War on Terror*. LSE International Studies. New York; London: Routledge, 2007.

Collins, Michael D. "A Fear of Flying: Diagnosing Traumatic Neurosis Among
British Aviators of the Great War." *First World War Studies* 6, no. 2 (2015):
187–202.

Cook, James L. "Review of *Killing Without Heart: Limits on Robotic Warfare in
an Age of Persistent Conflict*, by Shane M. Riza." *Journal of Military Ethics* 13
(2014): 106–111.

Coram, Robert. *Boyd: The Fighter Pilot Who Changed the Art of War*. Boston:
Little, Brown, 2002.

Correll, John T. "The Assault on EBO: The Cardinal Sin of Effects-Based Operations Was That It Threatened the Traditional Way of War." *Air Force
Magazine*, January 2013, 50–54.

Crowley, Michael, Falih Hassan, and Eric Schmitt. "U.S. Strike in Iraq Kills
Qassim Suleimani, Commander of Iranian Forces." *New York Times*, 7 January 2020.

Cunningham, Erin. "Iran's Gamble: Analysts Say Brazen Attack Aimed to
Pressure U.S. with Little Fear of Reprisal." *Washington Post*, 20 September
2019.

Daggett, Cara. "Drone Disorientations." *International Feminist Journal of Politics* 17, no. 3 (2015): 361–379.

Davenport, Thomas, and Ravi Kalakota. "The Potential for Artificial Intelligence in Healthcare." *Future Healthcare Journal* 6, no. 2 (2019): 94.

Defense Advanced Research Projects Agency (DARPA). "AlphaDogfight Trials Foreshadow Future of Human-Machine Symbiosis." DARPA, news release, 26 August 2020. www.darpa.mil/news-events/2020-08-26.

Defense Industry Daily Staff. "Tomahawk's Chops: xGM-109 Block IV Cruise Missiles." *Defense Industry Daily*, 13 July 2020.

de Peyer, Robin. "Sherlock Star Benedict Cumberbatch Saves Deliveroo Cyclist Being Violently Mugged by Gang of Thugs." *Evening Standard*, 2 June 2018.

de Visser, Ewart J., Samuel S. Monfort, Ryan McKendrick, Melissa A. B. Smith, Patrick E. McKnight, Frank Krueger, and Raja Parasuraman. "Almost Human: Anthropomorphism Increases Trust Resilience in Cognitive Agents." *Journal of Experimental Psychology: Applied* 22, no. 3 (2016): 331.

DeTeresa, Steven J., Michael W. McErlean, Noah B. Bleicher, Thomas E. Hawley, Sasha Rogers, and Lorry M. Fenner. *The Joint Improvised Explosive Device Defeat Organization: DoD's Fight Against IEDs Today and Tomorrow.* US House of Representatives, Committee on Armed Services, Subcommittee on Oversight & Investigations, November 2008. https://armedservices .house.gov/_cache/files/c/f/cfddccb2-fc15-4a3d-b7e3-50fe3ea68eca/D09F 0BEF55D1B39D2CC196408918781D.jieddo-report-11-08-vf.pdf.

Drescher, Kent David, David W. Foy, Caroline M. Kelly, Anna Leshner, Kerri Elizabeth Shutz, and Brett T. Litz. "An Exploration of the Viability and Usefulness of the Construct of Moral Injury in War Veterans." *Traumatology* 17, no. 1 (2011): 8–13.

Eberle, Christopher J. *Justice and the Just War Tradition: Human Worth, Moral Formation, and Armed Conflict.* New York: Routledge, 2016.

Echevarria, Antulio Joseph. *Clausewitz and Contemporary War.* Oxford: Oxford University Press, 2007.

Ehrhard, Thomas. *Air Force UAVs: The Secret History.* Edited by David A. Deptula. Arlington, VA: Mitchell Institute Press, 2010.

———. "Unmanned Aerial Vehicles in the United States Armed Services: A Comparative Study of Weapon System Innovation." Edited by Eliot A. Cohen. PhD diss., Johns Hopkins University, Baltimore, 2001.

Elias, Bart. *Unmanned Aircraft Operations in Domestic Airspace: U.S. Policy Perspectives and the Regulatory Landscape.* Congressional Research Service report 7-5700. Washington, DC, 27 January 2016. www.hsdl .org/?view&did=789767.

Enemark, Christian. *Armed Drones and the Ethics of War: Military Virtue in a Post-Heroic Age*. War, Conflict and Ethics. London: Routledge, 2014.

Epstein, Michael. "The Curious Case of Anwar Al-Aulaqi: Is Targeting a Terrorist for Execution by Drone Strike a Due Process Violation When the Terrorist Is a United States Citizen?" *Michigan State University College of Law Journal of International Law* 19 (2010): 723.

Evans, Christine. "Drones, Projections, and Ghosts: Restaging Virtual War in *Grounded* and *You Are Dead. You Are Here.*" *Theatre Journal* 67, no. 4 (2015): 663–686.

Fihn, Stephan D., Joseph Francis, Carolyn Clancy, Christopher Nielson, Karin Nelson, John Rumsfeld, Theresa Cullen, Jack Bates, and Gail L. Graham. "Insights from Advanced Analytics at the Veterans Health Administration." *Health Affairs* 33, no. 7 (2014): 1203–1211.

Fisher, Max. "Obama Finds Predator Drones Hilarious." *Atlantic*, 3 May 2010.

Foot, Philippa. *Virtues and Vices and Other Essays in Moral Philosophy*. Berkeley: University of California Press, 1978.

Franke, Ulrike Esther. "The Global Diffusion of Unmanned Aerial Vehicles (UAVs), or 'Drones.'" In *Precision Strike Warfare and International Intervention: Strategic, Ethico-Legal, and Decisional Implications*. Edited by Mike Aaronson, Wali Aslam, Tom Dyson, and Regina Rauxloh, 52–72. London: Routledge, 2015.

———. "The Unmanned Revolution: How Drones Are Revolutionising Warfare." PhD diss., University of Oxford, 2018.

Frankfurt, Sheila. "An Empirical Investigation of Moral Injury in Combat Veterans." PhD diss., University of Minnesota, 2015.

Fraser, Douglas M., and Wendell S. Hertzelle. "Haiti Relief: An International Effort Enabled Through Air, Space, and Cyberspace." *Air & Space Power Journal* 24, no. 4 (2010): 5–12.

French, Shannon E. *The Code of the Warrior: Exploring Warrior Values Past and Present*. 2nd ed. Lanham, Maryland: Rowman & Littlefield, 2017.

Frevert, Ute. *Men of Honour: A Social and Cultural History of the Duel*. Cambridge: Polity, 1995.

Galliott, Jai. *Military Robots: Mapping the Moral Landscape*. Military and Defence Ethics. Farnham, UK: Ashgate, 2015.

Geneva Convention. Additional Protocol I to the Geneva Conventions. (I) Article 24. 1977.

Gettinger, Dan, Arthur Holland Michel, Alex Pasternack, Jason Koebler, Shawn Musgrave, and Jared Rankin. *The Drone Primer: A Compendium of the Key Issues*. New York: Annandale-On-Hudson, 2014. https://drone center.bard.edu/the-drone-primer-announcement.

Gettinger, Dan, and Arthur Holland Michel. "Loitering Munitions." Center for the Study of the Drone, Bard College, 2017. https://dronecenter.bard .edu/files/2017/02/CSD-Loitering-Munitions.pdf.

Grant, Rebecca. "The ROVER." *Air Force Magazine*, August 2013.

Grant, Ulysses S. *The Papers of Ulysses S. Grant, Vol. 1, 1837–1861.* Carbondale: Southern Illinois University Press, 1967.

Gray, Colin S. "War: Continuity in Change, and Change in Continuity." *Parameters* 40, no. 2 (2010): 5.

Gray, J. Glenn. *The Warriors: Reflections on Men in Battle.* Lincoln: University of Nebraska Press, 1998.

Gregory, Derek. "From a View to a Kill." *Theory, Culture & Society* 28, no. 7–8 (2012): 188–215.

Grossman, Dave. *On Killing: The Psychological Cost of Learning to Kill in War and Society*, rev. ed. New York: Little, Brown, 2009.

Gusterson, Hugh. *Drone: Remote Control Warfare.* Cambridge, MA: MIT Press, 2015.

Hallion, Richard P. "Fighter Aircraft." In *The Oxford Companion to American Military History*, edited by Richard Holmes, Charles Singleton, and Spencer Jones. Oxford: Oxford University Press, 2000.

Hamilton, Alexander. "Letter to Edward, 11 November 1769." *The Papers of Alexander Hamilton: Digital Edition.* Vol. 1, *1768–1778.* Edited by Harold C. Syrett. Charlottesville: University of Virginia Press, Rotunda, 2011. http://rotunda.upress.virginia.edu/founders/default .xqy?keys=ARHN-print-01-01-02-0002.

Hampton, Dan. *Viper Pilot: The Autobiography of One of America's Most Decorated F-16 Combat Pilots.* New York: William Morrow, 2012.

Hanson, Victor Davis. *The Western Way of War: Infantry Battle in Classical Greece.* New York: A. A. Knopf, 1989.

Harris, Shane, Erin Cunningham, and Kareem Fahim. "Trump Stops Short of Directly Blaming Iran for Attack on Saudi Oil Facilities." *Washington Post*, 17 September 2019.

Harris, William V. *War and Imperialism in Republican Rome: 327–70 B.C.* Oxford and New York: Clarendon Press and Oxford University Press, 1979.

Harwell, Drew. "Google to Drop Pentagon AI Contract After Employee Objections to the 'Business of War.'" *Washington Post*, 1 June 2018.

Hastings, Max. "Gatwick Drone Shambles Is Just a Taste of What's to Come." *Times* (London), 21 December 2018.

———. *Overlord: D-Day and the Battle for Normandy.* London: Pan Books, 2015.

Hennigan, W. J. "Experts Say Drones Pose a National Security Threat—and We Aren't Ready." *Time*, 31 May 2018.

Hill, Ryan. "Have I Ever Been to War?" *Military Review*, January–February 2020.

Himes, Kenneth R. *Drones and the Ethics of Targeted Killing*. Lanham, MD: Rowman and Littlefield, 2016.

Holder, Eric. "Attorney General Eric Holder Speaks at Northwestern University School of Law." Chicago: Office of Public Affairs, 2012.

Holland, James. *The Battle of Britain: Five Months That Changed History, May–October 1940*. London: Corgi, 2011.

Holmes, Oliver Wendell. *Speeches*. Making of Modern Law. Boston: Little, Brown, 1918.

Homer. *Homer's Iliad*. Translated by George Chapman. MHRA Tudor & Stuart Translations. Cambridge: Modern Humanities Research Association, 2017.

———. *The Iliad of Homer*. Translated by Alexander Pope. Chadwyck-Healey Literature Collections. Cambridge: Chadwyck-Healey, 1992.

———. *The Iliad of Homer*. Translated by Ernest Myers, Walter Leaf, and Andrew Lang. New York: Modern Library, 1929.

———. *The Odyssey*. Translated by Alexander Pope. Doylestown, PA: Wildside Press, 2003.

Hubbard, Ben, Pako Karasz, and Stanley Reed. "Two Major Saudi Oil Installations Hit by Drone Strike, and U.S. Blames Iran." *New York Times*, 14 September 2019.

Human Rights Watch. "Iraq: ISIS Escapees Describe Systematic Rape." 14 April 2015. www.hrw.org/news/2015/04/14/iraq-isis-escapees-describe-systematic-rape.

Hursthouse, Rosalind. *On Virtue Ethics*. Oxford: Oxford University Press, 1999.

Insinna, Valerie. "In the Fight Against ISIS, Predators and Reapers Prove Close-Air Support Bona-Fides." *Defense News*, 8 August 2017.

"Irresistible Force." *Popular Mechanics*, February 1992, 102–104.

Israel Aerospace Industries. "Harpy: Autonomous Weapon for All Weather." www.iai.co.il/p/harpy.

Jackson, Jonathan L. "Solving the Problem of Time-Sensitive Targeting." Paper, Naval War College, 2003.

Jiang, Fei, Yong Jiang, Hui Zhi, Yi Dong, Hao Li, Sufeng Ma, Yilong Wang, Qiang Dong, Haipeng Shen, and Yongjun Wang. "Artificial Intelligence in Healthcare: Past, Present and Future." *Stroke and Vascular Neurology* 2, no. 4 (2017): 230–243.

Johnson, Lyndon B. "Address on Vietnam Before the National Legislative Conference, San Antonio, Texas." American Presidency Project, 29

September 1967. www.presidency.ucsb.edu/documents/address-vietnam
-before-the-national-legislative-conference-san-antonio-texas.

Johnson, Rebecca J. "The Wizard of Oz Goes to War: The Unmanned Systems
in Counterinsurgency." In *Killing by Remote Control: The Ethics of an Unmanned Military*, edited by Bradley Jay Strawser, 154–178. Oxford: Oxford
University Press, 2013.

Joint Chiefs of Staff. *Doctrine for the Armed Forces of the United States*, Joint
Publication 1. Washington, DC: Joint Staff, 2013.

———. *Close Air Support*, Joint Publication 3-09.3. Washington, DC: Joint
Staff, 2014.

———. *Counterterrorism*, Joint Publication 3-26. Washington, DC: Joint
Staff, 2013.

Junger, Sebastian. *War*. New York: Twelve, 2010.

Kaag, John J. "Drones, Ethics and the Armchair Soldier." *Opinionator* (blog).
*New York Times*, 17 March 2013.

Kaag, John J., and Sarah E. Kreps. *Drone Warfare*. Cambridge: Polity, 2014.

———. "The Moral Hazard of Drones." *New York Times*, 22 July 2012.

Kaurin, Pauline M. Shanks. *The Warrior, Military Ethics and Contemporary
Warfare: Achilles Goes Asymmetrical*. Military and Defence Ethics. Farnham,
UK: Ashgate, 2014.

———. "Healing the Wounds of War: Moral Luck, Moral Uncertainty, and
Moral Injury." *Strategy Bridge*, 5 January 2018.

Kean, Thomas H., Lee H. Hamilton, Richard Ben-Veniste, Bob Kerrey,
Fred F. Fielding, John F. Lehman, Jamie S. Gorelick, Timothy J. Roemer,
Slade Gorton, and James R. Thompson. *The 9/11 Commission Report*. National Commission on Terrorist Attacks upon the United States, 2004.
www.9-11commission.gov/report/911Report.pdf.

Kelly, Kevin. *What Technology Wants*. New York: Viking, 2010.

Kennebeck, Sonia, and Ines Hofmann Kanna. *National Bird*. Edited by Lois
Vossen. Independent Lens documentary film. Public Broadcasting System,
2017.

Kidwell, Deborah C. "The U.S. Experience: Operational." In *Precision and
Purpose: Airpower in the Libyan Civil War*, edited by Karl P. Meuller, 107–
152. Santa Monica, CA: RAND, 2015.

Killmister, Suzy. "Remote Weaponry: The Ethical Implications." *Journal of
Applied Philosophy* 25, no. 2 (2008): 121–133.

Kirkpatrick, Jesse. "Drones and the Martial Virtue Courage." *Journal of Military Ethics* 14, no. 3–4 (2015): 202–219.

———. "Reply to Sparrow: Martial Courage—or Merely Courage?" *Journal
of Military Ethics* 14, no. 3–4 (2015): 228–231.

Komarovskaya, Irina, Shira Maguen, Shannon E. McCaslin, Thomas J. Metzler, Anita Madan, Adam D. Brown, Isaac R. Galatzer-Levy, Clare Henn-Haase, and Charles R. Marmar. "The Impact of Killing and Injuring Others on Mental Health Symptoms Among Police Officers." *Journal of Psychiatric Research* 45, no. 10 (2011): 1332–1336.

Kozaryn, Linda D. "Fog, Friction Rule Takur Ghar Battle." American Forces Press Service, news release, 24 May 2002. www.globalsecurity.org/military /library/news/2002/05/mil-020524-dod02.htm.

Kreps, Sarah E. *Drones: What Everyone Needs to Know.* New York: Oxford University Press, 2016.

Lambeth, Benjamin S. *Air Power Against Terror: America's Conduct of Operation Enduring Freedom.* Santa Monica, CA: RAND, 2005.

———. "Airpower, Spacepower, and Cyberpower." *Joint Force Quarterly* 60, no. 1 (2011): 46–53.

———. *NATO's Air War for Kosovo: A Strategic and Operational Assessment.* Santa Monica, CA: RAND, 2001.

Lamothe, Dan. "Investigation: Friendly Fire Airstrike That Killed U.S. Special Forces Was Avoidable." *Washington Post,* 4 September 2014.

Larm, Dennis. "Expendable Remotely Piloted Vehicles for Strategic Offensive Airpower Roles." MPhil thesis, School of Advanced Airpower Studies, Maxwell Air Force Base, Alabama, 1996.

Lavalle, Arthur J. C., ed. *The Tale of Two Bridges and the Battle for the Skies over North Vietnam.* Vol. 1. Columbia, PA: Diane Publishing, 1976.

Lee, Caitlin. "The Culture of US Air Force Innovation: A Historical Case Study of the Predator Program." PhD diss., King's College London, 2016.

Lee, Peter. "The Distance Paradox: Reaper, the Human Dimension of Remote Warfare, and Future Challenges for the RAF." *Air Power Review* 21, no. 3 (2018): 106–130.

———. "Heroes and Cowards: Genealogy, Subjectivity and War in the Twenty-First Century." *Genealogy* 2, no. 2 (2018): 15.

———. *Reaper Force: The Inside Story of Britain's Drone Wars.* London: John Blake Publishing, 2018.

Lungescu, Oana, Giampaolo Di Paola, and Charles Brouchard. "Press Briefing." North Atlantic Treaty Organization, news release, 31 March 2011, www.nato.int/cps/en/natolive/opinions_71897.htm.

Lyman, Princeton N., and F. Stephen Morrison. "The Terrorist Threat in Africa." *Foreign Affairs* 83, no. 1 (2004): 75–86.

MacMillan, Douglas. "Google Won't Seek to Renew Pentagon Contract after Internal Backlash." *Wall Street Journal,* 1 June 2018.

MacNair, Rachel. *Perpetration-Induced Traumatic Stress: The Psychological Consequences of Killing.* Psychological Dimensions to War and Peace. London: Praeger, 2002.

Maggio, Edoardo. "Putin Believes That Whatever Country Has the Best AI Will Be 'the Ruler of the World.'" *Business Insider,* 4 September 2017.

Maguen, Shira, and Brett T. Litz. "Moral Injury in Veterans of War." *PTSD Research Quarterly* 23, no. 1 (2012): 1–6.

Maguen, Shira, Barbara A. Lucenko, Mark A. Reger, Gregory A. Gahm, Brett T. Litz, Karen H. Seal, Sara J. Knight, and Charles R. Marmar. "The Impact of Reported Direct and Indirect Killing on Mental Health Symptoms in Iraq War Veterans." *Journal of Traumatic Stress* 23, no. 1 (February 2010): 86–90.

Maguen, Shira, Thomas J. Metzler, Brett T. Litz, Karen H. Seal, Sara J. Knight, and Charles R. Marmar. "The Impact of Killing in War on Mental Health Symptoms and Related Functioning." *Journal of Traumatic Stress* 22, no. 5 (October 2009): 435–443.

Mark, Joyal, Iain McDougall, and J. C. Yardley. *Greek and Roman Education: A Sourcebook.* London: Routledge, 2009.

Marlantes, Karl. *What It Is Like to Go to War.* London: Corvus, 2011.

Marshall, Samuel Lyman Atwood. *Men Against Fire: The Problem of Battle Command in Future War.* Washington, DC: Infantry Journal Press, 1947.

Mattis, James N. *Summary of the National Defense Strategy.* Department of Defense, Washington, DC, 2018.

———. "USJFCOM Commander's Guidance for Effects-Based Operations." *Joint Force Quarterly* 4, no. 51 (2008).

Mayer, Jane. "The Predator War." *New Yorker* 26 (2009): 36–45.

McCarthy, Jacob R. "Tried and True to Air Force Blue: A Leader Remembered." Nellis Air Force Base, news release, 30 August, 2007, www .nellis.af.mil/News/Article/285977/tried-and-true-to-air-force-blue-a-leader -remembered.

McDonnell, John P. "Apportion or Divert? The JFC's Dilemma: Asset Availability for Time-Sensitive Targeting." Thesis, Naval War College, 2002.

McFarland, Stephen L. *A Concise History of the U.S. Air Force.* Washington DC: Air Force History and Museums Program, 1997.

McInerny, Ralph, and John O'Callaghan. "Saint Thomas Aquinas." In *The Stanford Encyclopedia of Philosophy,* edited by Edward N. Zalta. Stanford, CA: Stanford University, 2018.

Menthe, Lance, Amado Cordova, Carl Rhodes, Rachel Costello, and Jeffrey Sullivan. *The Future of Air Force Motion Imagery Exploitation: Lessons from the Commercial World.* Santa Monica, CA: RAND, 2012. www.rand.org /content/dam/rand/pubs/technical_reports/2012/RAND_TR1133.pdf.

Mewett, Christopher. "Understanding War's Enduring Nature Alongside Its Changing Character." *War on the Rocks*, 21 January 2014.

Miller, Donald L. *Masters of the Air: America's Bomber Boys Who Fought the Air War Against Nazi Germany*. New York: Simon & Schuster, 2006.

Miller, Grace E. "'Boom / [S]He Is Not': Drone Wars and the Vanishing Pilot." *War, Literature and the Arts* 29 (2017): 1–17.

Mindell, David A. *Our Robots, Ourselves: Robotics and the Myths of Autonomy*. New York: Viking Adult, 2015.

Missile Defense Project. "Brimstone." Center for Strategic and International Studies, 6 December 2017; updated 30 July 2021. https://missilethreat.csis .org/missile/brimstone.

———. "JASSM / JASSM-ER (AGM-158A/B)." Center for Strategic and International Studies, 6 October 2016; updated 30 July 2021. https://missile threat.csis.org/missile/jassm.

Mizokami, Kyle. "Air Force Tests New 'Loyal Wingman' Sidekick Drone for Combat." *Popular Mechanics*, 7 March 2019.

Moelker, René, and Peter Olsthoorn. "Virtue Ethics and Military Ethics." *Journal of Military Ethics* 6, no. 4 (2007): 257–258.

Molendijk, Tine, Eric-Hans Kramer, and Désirée Verweij. "Moral Aspects of 'Moral Injury': Analyzing Conceptualizations on the Role of Morality in Military Trauma." *Journal of Military Ethics* 17, no. 1 (2018): 36–53.

Moschgat, James E. *The Composite Wing: Back to the Future*. Maxwell AFB, AL: Air University Press, 1992.

Mueller, John. "The Perfect Enemy: Assessing the Gulf War." *Security Studies* 5, no. 1 (1995): 77–117.

Nash, William P., Teresa L. Marino Carper, Mary Alice Mills, Teresa Au, Abigail Goldsmith, and Brett T. Litz. "Psychometric Evaluation of the Moral Injury Events Scale." *Military Medicine* 178, no. 6 (June 2013): 646–652.

Naylor, Sean. *Not a Good Day to Die: The Untold Story of Operation Anaconda*, rev. ed. London: Penguin, 2006.

Nolan, Cathal J. *The Allure of Battle: A History of How Wars Have Been Won and Lost*. New York: Oxford University Press, 2019.

Noordally, Ryan. "On the Toxicity of the 'Warrior' Ethos." *Wavell Room: Contemporary British Military Thought*, 28 April 2020.

Office of Legal Counsel. Memorandum for the Attorney General Re: Applicability of Federal Criminal Laws and the Constitution to Contemplated Lethal Operations against Shaykh Anwar Al-Aulaqi. Washington, DC, 2010.

Olsthoorn, Peter. "Courage in the Military: Physical and Moral." *Journal of Military Ethics* 6, no. 4 (13 December 2007): 270–279.

——. "Honor as a Motive for Making Sacrifices." *Journal of Military Ethics* 4, no. 3 (1 November 2005): 183–197.

——. *Military Ethics and Virtues: An Interdisciplinary Approach for the 21st Century.* London and New York: Routledge, 2011.

Pallardy, Richard. "2010 Haiti Earthquake." In *Encyclopædia Britannica*. Chicago: Encyclopædia Britannica, 2020.

Pellerin, Cheryl. "Project Maven to Deploy Computer Algorithms to War Zone by Year's End." US Department of Defense, news release, 21 July 2017. www.defense.gov/Explore/News/Article/Article/1254719/project-maven-to-deploy-computer-algorithms-to-war-zone-by-years-end.

Petrenko, Anton. "Between Berserksgang and the Autonomous Weapons Systems." *Public Affairs Quarterly* 26, no. 2 (2012): 81–102.

Plato. "Laches." In *The Collected Dialogues of Plato*, edited by Edith Hamilton. Bollingen Series, 71:123–144. Princeton, NJ: Princeton University Press, 2002.

——. "The Republic." In *The Collected Dialogues of Plato*, edited by Edith Hamilton. Bollingen Series, 71:575–844. Princeton, NJ: Princeton University Press, 2002.

Power, Matthew. "Confessions of a Drone Warrior." *GQ*, 23 October 2013.

Powers, Marina. "Sticks and Stones: The Relationship Between Drone Strikes and Al-Qaeda's Portrayal of the United States." *Critical Studies on Terrorism* 7, no. 3 (2 September 2014): 411–421.

Press, Eyal. "The Wounds of the Drone Warrior." *New York Times Magazine*, 13 July 2018.

Pruitt, Sarah. "Heroes of Pearl Harbor: George Welch and Kenneth Taylor." *History*, 28 November 2016; updated 7 December 2018. www.history.com/news/heroes-of-pearl-harbor-george-welch-and-kenneth-taylor.

Rae, James DeShaw, and John T. Crist. *Analyzing the Drone Debates: Targeted Killing, Remote Warfare, and Military Technology.* Palgrave Pivot. Basingstoke, UK: Palgrave Macmillan, 2014.

Reagan, Ronald. "Trancript of Address by Reagan on Libya." *New York Times*, 15 April 1986.

Renic, Neil C. *Asymmetric Killing: Risk Avoidance, Just War, and the Warrior Ethos.* Oxford: Oxford University Press, 2020.

——. "UAVs and the End of Heroism? Historicising the Ethical Challenge of Asymmetric Violence." *Journal of Military Ethics* 17, no. 4 (2018): 188–197.

"Retired Gen. Creech, 'Father of the Thunderbirds,' Dies." *Las Vegas Sun*, 28 August 2003.

Riza, M. Shane. *Killing Without Heart: Limits on Robotic Warfare in an Age of Persistent Conflict.* Washington, DC: Potomac Books, 2013.

———. "Two-Dimensional Warfare: Combatants, Warriors, and Our Post-Predator Collective Experience." *Journal of Military Ethics* 13, no. 3 (2014): 257–273.

Roberts, Huw, Josh Cowls, Jessica Morley, Mariarosaria Taddeo, Vincent Wang, and Luciano Floridi. "The Chinese Approach to Artificial Intelligence: An Analysis of Policy, Ethics, and Regulation." *AI & Society* 36, no. 1 (1 March 2021): 59–77.

Robertson, Linda R. *The Dream of Civilized Warfare: World War I Flying Aces and the American Imagination.* Minneapolis: University of Minnesota Press, 2003.

Robinson, Paul. "Magnanimity and Integrity as Military Virtues." *Journal of Military Ethics* 6, no. 4 (1 December 2007): 259–269.

Roby, Cheryl J., and David S. Alberts. *NATO NEC C2 Maturity Model.* Washington, DC: Center for Advanced Concepts and Technology (ACT), 2010. https://apps.dtic.mil/dtic/tr/fulltext/u2/a555717.pdf.

Roosevelt, Franklin D. "Speech by Franklin D. Roosevelt, New York (Transcript)." Address to Congress, 8 December 1941. Library of Congress, http://hdl.loc.gov/loc.afc/afc1986022.ms2201.

Rosen, Brianna. "To End the Forever Wars, Rein in the Drones." *Just Security,* 16 February 2021.

Rosenberg, Matthew. "Pentagon Details Chain of Errors in Strike on Afghan Hospital." *New York Times,* 29 April 2016.

Rostker, Bernard D., Charles Nemfakos, Henry A. Leonard, Elliot Axelband, Abby Doll, Kimberly N. Hale, Brian McInnis, Richard Mesic, Daniel Tremblay, Roland J. Yardley, and Stephanie Young. *Building Toward an Unmanned Aircraft System Training Strategy.* Santa Monica, CA: RAND, 2014. https://apps.dtic.mil/dtic/tr/fulltext/u2/a607338.pdf.

Royakkers, L., and R. van Est. "The Cubicle Warrior: The Marionette of Digitalized Warfare." *Ethics and Information Technology* 12, no. 3 (September 2010): 289–296.

Salem, Maha, Friederike Eyssel, Katharina Rohlfing, Stefan Kopp, and Frank Joublin. "Effects of Gesture on the Perception of Psychological Anthropomorphism: A Case Study with a Humanoid Robot." Paper presented at the International Conference on Social Robotics, Berlin, 2011.

Sauer, Frank, and Niklas Schörnig. "Killer Drones: The 'Silver Bullet' of Democratic Warfare?" *Security Dialogue* 43, no. 4 (2012): 363–380.

Savage, Charlie. "Trump Revokes Obama-Era Rule on Disclosing Civilian Casualties from U.S. Airstrikes Outside War Zones." *New York Times,* 6 March 2019.

Savage, Charlie, and Eric Schmitt. "Biden Secretly Limits Counterterrorism Drone Strikes Away from War Zones." *New York Times,* 3 March 2021.

Scaparrotti, Curtis M., and Denis Mercier. *Framework for Future Alliance Operations.* North Atlantic Treaty Organization, 2018. www.act.nato.int /images/stories/media/doclibrary/180514_ffao18.pdf.

Scarborough, Rowan. "Pentagon Uproar over Panetta's Hero Medals for Drone Operators, Cybersleuths." *Washington Times,* 15 February 2013.

Schilling, Dan, and Lori Chapman Longfritz. *Alone at Dawn.* New York: Grand Central Publishing, 2019.

Schlight, John. "Project CHECO Southeast Asia Report: Jet Forward Air Controllers in SEAsia." Hickam Air Force Base, HI: Department of the Air Force, 1969.

Schmidle, Nicholas. "Getting Bin Laden." *New Yorker,* 1 August 2011.

Schmitt, Eric. "A Botched Drone Strike in Kabul Started with the Wrong Car." *New York Times,* 21 September 2021.

Schroeder, Juliana, and Matthew Schroeder. "Trusting in Machines: How Mode of Interaction Affects Willingness to Share Personal Information with Machines." Proceedings of the 51st Hawaii International Conference on System Sciences. *Social and Psychological Perspectives in Collaboration Research,* 3 January 2018.

Schultz, Timothy P. *The Problem with Pilots: How Physicians, Engineers, and Airpower Enthusiasts Redefined Flight.* Baltimore: Johns Hopkins University Press, 2018.

Schulzke, Marcus. "Rethinking Military Virtue Ethics in an Age of Unmanned Weapons." *Journal of Military Ethics* 15, no. 3 (1 July 2016): 187–204.

Schwarz, Elke. "Written Submission of Evidence to the All Party Parliamentary Group (APPG) on Drones: Ethical Challenges." All Party Parliamentary Group on Drones, December 2017. http://appgdrones.org.uk/wp -content/uploads/2014/08/10.-Elke-APPG-Drones-041217.pdf.

Schwartz, Matthew S. "Who Was the Iraqi Commander Also Killed in the Baghdad Drone Strike?" *NPR,* 4 January 2020.

Sewall, Sarah B. *Chasing Success: Air Force Efforts to Reduce Civilian Harm.* Maxwell Air Force Base, AL: Air University Press, 2016. www.airuniversity .af.edu/Portals/10/AUPress/Books/B_0142_SEWALL_CHASING _SUCCESS.pdf.

Shane, Scott. *Objective Troy: A Terrorist, a President, and the Rise of the Drone.* New York: Seal Books, 2016.

Shane, Scott, and Daisuke Wakabayashi. "'The Business of War': Google Employees Protest Work for the Pentagon." *New York Times,* 4 April 2018.

Shay, Jonathan. *Achilles in Vietnam: Combat Trauma and the Undoing of Character,* trade pbk. ed. New York: Scribner, 2003.

Shea, Jamie. "Precision Strike Capabilities: Political and Strategic Consequences." Preface to *Precision Strike Warfare and International Intervention:*

*Strategic, Ethico-Legal, and Decisional Implications*, edited by Mike Aaronson, Wali Aslam, Tom Dyson, and Regina Rauxloh. London: Routledge, 2015.

Shea, Tim. "A Combat Badge Does Not a Soldier Make." In *The Angry Staff Officer* (blog), 1 February 2016. https://angrystaffofficer.com/2016/02/01/a-combat-badge-does-not-a-soldier-make.

Shepherd, Jack. "Benedict Cumberbatch Hailed a 'Hero' After Fending Off Four Muggers Attacking a Deliveroo Cyclist." *Independent* (London), 2 June 2018.

Sherman, Nancy. *Afterwar: Healing the Moral Injuries of Our Soldiers.* New York: Oxford University Press, 2015.

Sime, Ruth Lewin. *Lise Meitner: A Life in Physics.* California Studies in the History of Science, vol. 13. Berkeley: University of California Press, 1996.

Sirak, Michael C. "ISR Revolution." *Air Force Magazine,* 1 June 2010, 36–42.

Skowronski, Will. "Reapers and the RPA Resurgence: The MQ-9 Can Perform Strike, Coordination and Reconnaissance Against High-Value Targets." *Air Force Magazine*, August 2016.

Sparrow, Rob. "War Without Virtue?" In *Killing by Remote Control*, edited by Bradley Jay Strawser, 84–105. Oxford: Oxford University Press, 2013.

———. "Martial and Moral Courage in Tele-operated Warfare: A Commentary on Kirkpatrick." *Journal of Military Ethics* 14, no. 3–4 (2015): 220–227.

Storr, Jim. "A Command Philosophy for the Information Age: The Continuing Relevance of Mission Command." *Defence Studies* 3, no. 3 (2003): 119–129.

Strawser, Bradley Jay. "Moral Predators: The Duty to Employ Uninhabited Aerial Vehicles." *Journal of Military Ethics* 9, no. 4 (2010): 342–368.

Swinford, Steven. "Osama Bin Laden Dead: Blackout During Raid on Bin Laden Compound." *Telegraph* (London), 4 May 2011.

Szoldra, Paul. "Watch John Chapman's Incredible Heroics in the First Medal of Honor Action Ever Recorded on Video." *Task and Purpose*, 10 July 2019.

Telegraph Reporters. "Benedict Cumberbatch Fights Off Four Muggers Who Attacked Deliveroo Cyclist near Baker Street." *Telegraph* (London), 1 June 2018.

Thompson, James. "Phantom of Takur Ghar: The Predator Above Roberts Ridge." US Air Force, Air Combat Command, news release, 30 August 2018. www.acc.af.mil/News/Article-Display/Article/1617739/phantom-of-takur-ghar-the-predator-above-roberts-ridge.

Thornhill, Paula G. *"Over Not Through": The Search for a Strong, Unified Culture for America's Airmen.* Santa Monica, CA: RAND, 2012.

Tierney, Dominic. "The Twenty Years' War." *Atlantic*, 23 August 2016.

Tirpak, John A. "Bombers over Libya." *Air Force Magazine*, July 2011, 36–39.

Tirpak, John A., and Brian Everstine. "Syria Strike Marks Combat Debut for JASSM-ER." *Air Force Magazine*, 15 April 2018.

Turak, Natasha. "Pentagon Is Scrambling as China 'Sells the Hell out of' Armed Drones to US Allies." *CNBC*, 21 February 2019.

UK Ministry of Defence. *Future of Command and Control*. Swindon, UK: Development, Concepts and Doctrine Centre, 2017.

United Nations. "Charter of the United Nations." *Yale Law Journal* 55, no. 5 (1946): 1291–1317.

United Press International. "4 Pilots Killed as Stunt Planes Crash in Desert." *New York Times*, 19 January 1982.

———. "Air Force Finds Mechanical Failure Led to Crashes of Flying Team." *New York Times*, 11 April 1982.

US Air Force. *Air Force Instruction 11-2C-17*. Vol. 3, *Operations Procedures*. Washington, DC: USAF, 2015.

———. *Air Force Instruction 11-202*. Vol. 3, *General Flight Rules*. Washington, DC: USAF, 2019.

———. *Air Force Instruction 36-2643, Air Force Mentoring Program*. Washington, DC: USAF, 2019.

———. *Annex 3-0: Operations and Planning*. Maxwell Air Force Base, AL: LeMay Center for Doctrine Development and Education, 2016.

———. *Annex 3-03: Counterland Operations*. Maxwell Air Force Base, AL: LeMay Center for Doctrine Development and Education, 2019.

———. *Annex 3-60: Targeting*. Maxwell Air Force Base, AL: LeMay Center for Doctrine Development and Education, 2019.

———. *Air Force Instruction 13-112*. Vol. 1, *Joint Terminal Attack Controller (JTAC) Training*. Washington, DC: USAF, 2017.

———. "History of Creech Air Force Base." US Air Force, fact sheet, 16 May 2013. www.creech.af.mil/About-Us/Fact-Sheets/Display/Article/449127 /history-of-creech-air-force-base.

US Air Force Research Lab (AFRL). "Skyborg: Open . . . Resilient . . . Autonomous." AFRL, 2020. https://afresearchlab.com/technology/vanguards /successstories/skyborg.

US Army. "Army Regulation 600-8-22: Military Awards." 2019.

US Army, Arlington National Cemetery. "Audie Murphy." www.arlington cemetery.mil/Explore/Notable-Graves/Medal-of-Honor-Recipients /World-War-II-MoH-recipients/Audie-Murphy.

US Army Aviation and Missile Life Cycle Management Command (AM-COM). "Hellfire." AMCOM, information page, n.d. https://history .redstone.army.mil/miss-hellfire.html.

US Congress. "Loss of Nationality by Native-Born or Naturalized Citizen; Voluntary Action; Burden of Proof; Presumptions." 8 US Code § 1481. Ithaca, NY: Cornell Law School, 1952.

———. National Defense Authorization Act for Fiscal Year 2012. 112th Congress, Public Law 112-81.

———. Public Law 107-40 (Joint Resolution to Authorize the Use of United States Armed Forces Against Those Responsible for the Recent Attacks Launched Against the United States). 107th Congress, 2001.

US Department of Defense (DoD). "DoD Adopts Ethical Principles for Artificial Intelligence." DoD, news release, 24 February 2020. www.defense.gov/Newsroom/Releases/Release/Article/2091996/dod-adopts-ethical-principles-for-artificial-intelligence.

———. "Summary of the 2018 Department of Defense Artificial Intelligence Strategy: Harnessing AI to Advance Our Security and Prosperity." DoD, Washington, DC, 8 November 2018.

———. Unmanned Aerial Vehicles Roadmap. Washington, DC: DoD, 2000.

US House of Representatives, Committee on Armed Services. Effective Counterinsurgency: The Future of the U.S.-Pakistan Military Partnership. Hearing held 23 April 2009. Washington, DC: US Government Printing Office, 2010.

US Navy. "Tomahawk Cruise Missile." US Navy Office of Information, news release, 27 September 2021. www.navy.mil/Resources/Fact-Files/Display-FactFiles/Article/2169229/tomahawk-cruise-missile.

US Space Force. Spacepower: Doctrine for Space Forces. Space Capstone Publication (Washington, DC: US Space Force, 2020).

Van Dyk, Jere. "Who Were the 4 U.S. Citizens Killed in Drone Strikes?" CBS News, 23 May 2013.

von Luck, Hans. "The End in North Africa." In Experience of War: An Anthology of Articles from MHQ, the Quarterly Journal of Military History, edited by Robert Cowley, 430–442. New York: W. W. Norton and Company, 1992.

Waddington, Conway. "Drones: Degrading Moral Thresholds for the Use of Force and the Calculations of Proportionality." In Precision Strike Warfare and International Intervention: Strategic, Ethico-Legal, and Decisional Implications, edited by Mike Aaronson, Wali Aslam, Tom Dyson, and Regina Rauxloh, 114–151. London: Routledge, 2015.

Walker, Lauren. "Death from Above: Confessions of a Killer Drone Operator." Newsweek, 19 November 2015.

Warfare Branch Editor, Land Operations. Warminster, UK: Land Warfare Development Centre, 2017.

Waytz, Adam, Joy Heafner, and Nicholas Epley. "The Mind in the Machine: Anthropomorphism Increases Trust in an Autonomous Vehicle." *Journal of Experimental Social Psychology* 52 (2014): 113–117.

Wells, Mark K. *Courage and Air Warfare: The Allied Aircrew Experience in the Second World War.* Cass Series—Studies in Air Power. London: Frank Cass, 1995.

Wertheimer, Roger. *Empowering Our Military Conscience: Transforming Just War Theory and Military Moral Education.* Military and Defence Ethics. Farnham, UK: Ashgate, 2010.

White, Dana W., and Kenneth F. McKenzie Jr. "Department of Defense Press Briefing by Pentagon Chief Spokesperson Dana W. White and Joint Staff Director Lt. Gen. Kenneth F. McKenzie Jr. in the Pentagon Briefing Room." DoD, news release, 14 April 2018, www.defense.gov /Newsroom/Transcripts/Transcript/Article/1493749/department-of -defense-press-briefing-by-pentagon-chief-spokesperson-dana-w-whit.

Whittle, Richard. *Predator: The Secret Origins of the Drone Revolution.* New York: Henry Holt and Company, 2014.

Williams, Brian Glyn. *Counter Jihad: America's Military Experience in Afghanistan, Iraq, and Syria.* Philadelphia: University of Pennsylvania Press, 2017.

Wilson, Gregory. "A Time-Critical Targeting Roadmap." Thesis, Air Command and Staff College, Maxwell Air Force Base, AL, 2002.

Winters, Richard D., and Cole C. Kingseed. *Beyond Band of Brothers: The War Memoirs of Major Dick Winters.* New York: Penguin, 2008.

Wolfendale, Jessica. "What Is the Point of Teaching Ethics in the Military?" In *Ethics Education in the Military*, edited by Paul Robinson, Nigel de Lee, and Don Carrick. New York: Ashgate, 2008.

Woods, Chris. *Sudden Justice: America's Secret Drone Wars.* New York: Oxford University Press, 2015.

Worden, R. Michael. *Rise of the Fighter Generals: The Problem of Air Force Leadership, 1945–1982.* Maxwell Air Force Base, AL: Air University Press, 1998.

Work, Robert. "Establishment of an Algorithmic Warfare Cross-Functional Team (Project Maven)." Memorandum of the Deputy Secretary of Defense for Secretaries of the Military Departments et al. Washington, DC, 26 April 2017. www.govexec.com/media/gbc/docs/pdfs_edit/establishment_of _the_awcft_project_maven.pdf.

Wright, John D. *The Language of the Civil War.* Westport, CT: Oryx Press, 2001.

Wright, Lawrence. *The Looming Tower: Al-Qaeda's Road to 9/11.* London: Penguin, 2011.

Wright, Orville. "Letter to C. M. Hitchcock, 1917." Flight's Future, Smith-
    sonian Education, web page. www.smithsonianeducation.org/educators
    /lesson_plans/wright/flights_future.html.
Yuhas, Alan. "Airstrike That Killed Suleimani Also Killed Powerful Iraqi Mili-
    tia Leader." *New York Times*, 3 January 2020.
Zabecki, David T., ed. *Germany at War: 400 Years of Military History*. 4 vols.
    Santa Barbara, CA: ABC-CLIO, LLC, 2014.
Zenko, Micah, and Sarah Kreps. *Limiting Armed Drone Proliferation*. New
    York: Council on Foreign Relations, 2014.

# INDEX

**Joseph O. Chapa** is a lieutenant colonel in the US Air Force. He has served as a Predator pilot and instructor pilot, an instructor of philosophy at the US Air Force Academy, and a staff officer at the Pentagon. He holds a PhD in philosophy from the University of Oxford. His work on military ethics has been published in several online, print, and peer-reviewed publications. He currently serves on the Air Staff, where he focuses on AI ethics for the Department of the Air Force.

PublicAffairs is a publishing house founded in 1997. It is a tribute to the standards, values, and flair of three persons who have served as mentors to countless reporters, writers, editors, and book people of all kinds, including me.

I. F. STONE, proprietor of *I. F. Stone's Weekly*, combined a commitment to the First Amendment with entrepreneurial zeal and reporting skill and became one of the great independent journalists in American history. At the age of eighty, Izzy published *The Trial of Socrates*, which was a national bestseller. He wrote the book after he taught himself ancient Greek.

BENJAMIN C. BRADLEE was for nearly thirty years the charismatic editorial leader of *The Washington Post*. It was Ben who gave the *Post* the range and courage to pursue such historic issues as Watergate. He supported his reporters with a tenacity that made them fearless and it is no accident that so many became authors of influential, best-selling books.

ROBERT L. BERNSTEIN, the chief executive of Random House for more than a quarter century, guided one of the nation's premier publishing houses. Bob was personally responsible for many books of political dissent and argument that challenged tyranny around the globe. He is also the founder and longtime chair of Human Rights Watch, one of the most respected human rights organizations in the world.

•     •     •

For fifty years, the banner of Public Affairs Press was carried by its owner Morris B. Schnapper, who published Gandhi, Nasser, Toynbee, Truman, and about 1,500 other authors. In 1983, Schnapper was described by *The Washington Post* as "a redoubtable gadfly." His legacy will endure in the books to come.

Peter Osnos, *Founder*